T0260207

ANNALS OF MATHEMATICS STUDIES
Number 18

An essay toward

A UNIFIED THEORY OF SPECIAL FUNCTIONS

based upon the functional equation

$$\frac{\partial}{\partial z} F(z, \alpha) = F(z, \alpha+1)$$

By

C. TRUESDELL

PRINCETON
PRINCETON UNIVERSITY PRESS
LONDON: GEOFFREY CUMBERLEGE
OXFORD UNIVERSITY PRESS
1948

Photo-Lithoprint Reproduction
EDWARDS BROTHERS, INC.
Lithoprinters
ANN ARBOR, MICHIGAN
1948

In Memoriam H. Bateman

PREFACE

Nature of This Essay.

This essay presents and explains one particular branch of analysis, specially developed to provide a new method of deducing and proving formal relations among special functions, a new method which turns out to be very powerful. In no sense an expository work, it does not aim to compile information on special functions or to outline known methods of treating them. The scattered and partial results obtained by previous students of differential-difference equations of the type studied in this essay are summarized in the historical note at the end (§20).

The essay is written from the point of view of a mathematician. Readers interested only in practical use of special functions are directed to Chapter I, §3, §4, §5 of Chapter II, Corollary [9.4] and the remarks following it in Chapter III, Chapter IV, and Chapter V.

Appendices.

Material well known to many readers or otherwise unsuitable for the body of the essay has been relegated to three appendices.

Enumeration.

Formulas are numbered within each section. In references to formulas of other sections the number of the section is prefixed to the number of the formula, e.g., formula (11.6) is the formula numbered (6) in §11. Capital Latin letters refer to formulas in Appendix I, e.g., formula (C.3) is the formula numbered (3) in part C of Appendix I.

References.

References in square brackets refer to the list at the end of the essay.

Circumstances of Composition.

I began the studies leading to this essay through encountering a special case [Truesdell 1]. In Ann Arbor in the summer of 1944 I proved limited forms of the existence and uniqueness theorems of §9 and discovered the essential reduction in §3. The main body of results in §5-§8, §10-§17, and §19 I worked out between December 1944 and September 1945 in leisure hours while a Staff Member at the Radiation Laboratory, Massachusetts Institute of Technology. In the summer of 1945 the late Professor Bateman read the manuscript and supplied a number of references, through which I became acquainted with the previous results of Appell, Bruwier, and Doetsch discussed in §20. Some of the material in §3-§5, §11-§16 and §19 was presented to the American Mathematical Society September 17, 1945 (Bull. Am. Math. Soc. 51 (1945) p. 883, Abstract No. 229). In leisure hours while employed at the Naval Ordnance Laboratory I made extensive revisions and added §18 in June 1946; later I put §9-§10 into their present form and presented them to the American Mathematical Society December 27, 1946 (Bull. Am. Math. Soc. 53 (1947), p. 59, Abstract No. 57). Finally a summary of some of the contents of this essay was published in the Procedings of the National Academy of Sciences [Truesdell 2].

Acknowledgement.

I wish to express my obligation to the late Professor Bateman and to thank Professors B. Friedman and W. Hurewicz and Dr. Max M. Munk for reading and criticizing portions of this essay. Miss Charlotte Brudno and Mrs. Peggy Matheny have kindly assisted in the preparation of the final manuscript.

C.A.T.

Naval Ordnance Laboratory, White Oak, Silver Spring, Md.
and
University of Maryland, College Park, Maryland.
May 7, 1947.

TABLE OF CONTENTS

Chapter I

THE OBJECT AND PLAN OF THIS ESSAY

§1. THE GENERAL PROBLEM

While the familiar transcendents of mathematical physics belong to classes of functions which have been studied in various penetrating investigations of an analytic or of an algebraic character, there remain many simple formal relations which as yet seem to constitute particular properties of the particular functions, and are subsumed by no general framework. My aim is to present a theory on the basis of which some of these apparently individual formulas will easily appear as natural consequences of a few general theorems.

Let us first examine some of the commonest of these relations and formulate some natural questions about them.

Generating Functions:

$$\exp \frac{1}{2} y(w - \frac{1}{w}) = \sum_{-\infty}^{\infty} w^n J_n(y), \tag{1}$$

$$(1 - 2wy + y^2)^{-\frac{1}{2}} = \sum_{n=0}^{\infty} w^n P_n(y), \tag{2}$$

$$(1 - w)^{-a-1} \exp \frac{-yw}{1-w} = \sum_{n=0}^{\infty} w^n L_n^{(a)}(w), \tag{3}$$

$$\exp(2wy - y^2) = \sum_{n=0}^{\infty} \frac{w^n}{n!} H_n(y). \tag{4}$$

There is no reason to consider these expansions the natural generators for the special functions concerned, simply because they are the most commonly found. The less familiar expansions

$$(5) \qquad J_0(\sqrt{y^2 - 2wy}) = \sum_{n=0}^{\infty} \frac{w^n}{n!} J_n(y),$$

$$(6) \qquad e^{wy} J_0(w\sqrt{1 - y^2}) = \sum_{n=0}^{\infty} \frac{w^n}{n!} P_n(y),$$

$$(7) \qquad I_0(2w\sqrt{y - 1})\, I_0(2w\sqrt{y + 1}) = \sum_{n=0}^{\infty} \frac{w^n}{(n!)^2} P_n(y),$$

$$(8) \qquad e^w (yw)^{-\frac{1}{2}a} J_a(2\sqrt{wy}) = \sum_{n=0}^{\infty} \frac{w^n}{\Gamma(a+n+1)} L_n^{(a)}(y),$$

$$(9) \qquad \frac{(w-1)^m e^{wy}}{m!} = \sum_{n=-m}^{\infty} y^m L_m^{(n)}(y) \frac{w^{n+m}}{(n+m)!},$$

are perhaps equally natural. Any classification of relations such as the series (1) to (9) becomes possible only within the framework of a particular theory; the expansion

$$(A) \qquad G(w, y) = \sum_{n=0}^{\infty} a_n w^n f_n(y)$$

is too general to be the profitable subject of an investigation aiming to discover results of formal interest. The functions $f_n(y)$ must be limited to a restricted class, such as the class of solutions of a particular equation. There is no reason for considering the series (1) as necessarily an analogue of the series (2), or the series (4) an analogue of the series (6), since not $P_n(y)$ but possibly $n!P_n(y)$, or $P_n(y)/n!$ or some more complicated expression involving $P_n(y)$, may be a member of our selected class of functions. Supposing a particular function $f_n(y)$ to be a member of this class, yet to be defined, the first question reasonably raised is:

Question 1. Can we evaluate the sum $G(t, x)$ in the formula (A), the a_n being given?

There is one case in which the expansion we have written down subordinates itself readily to a more general one: formula (5) is a special case of the formula

$$(10) \qquad (z+w)^{-\frac{1}{2}\alpha} J_\alpha(2\sqrt{z+w}) = \sum_{n=0}^{\infty} \frac{(-w)^n}{n!} z^{-\frac{\alpha+n}{2}} J_{\alpha+n}(2\sqrt{z}).$$

It is therefore natural to ask:

Question 2. Can we evaluate the function $H_b(w, y)$ defined by the series

$$(B) \qquad H_b(w, y) = \sum_{n=0}^{\infty} a_n w^n f_{n+b}(y),$$

the a_n being given?

Question 2 includes Question 1, of course; its answer would enable us to generalize the remainder of the formulas (1) to (9) in the way that the formula (10) generalizes the formula (5).

<u>n-th Derivative Formulas</u>:

$$(11) \qquad J_n(y) = (-)^n \left(\frac{y}{2}\right)^n \frac{d^n}{d\left(\frac{y^2}{4}\right)^n} J_0(y),$$

$$(12) \qquad P_n(y) = \frac{1}{2^n n!} \frac{d^n}{dy^n} (y^2 - 1)^n,$$

$$(13) \qquad H_n(y) = (-)^n e^{y^2} \frac{d^n}{dy^n} e^{-y^2},$$

$$(14) \qquad L_n^{(a)}(y) = \frac{e^y y^{-a}}{n!} \frac{d^n}{dy^n} (e^{-y} y^{n+a}).$$

There is no reason to suppose these formulas natural analogues of each other; the mere fact that the formula

(15) $$L_n^{(m)}(y) = \frac{(-)^m}{n!} e^y \frac{d^{m+n}}{dy^{m+n}} (e^{-y} y^m)$$

is of more restricted validity than the formula (14), or that the formula

(16) $$P_n(\cos \theta) = \frac{(-)^n}{n!} \csc^{n+1} \theta \frac{d^n \sin \theta}{d(\cot \theta)^n}$$

is less familiar and less useful than the formula (12), does not make them any less likely to be the proper parallel cases to formulas (11) and (13) respectively. To obtain a natural classification we need the answer to:

Question 3. Do there exist functions $g(y, n)$, $q(y, n)$, and $h(y)$ such that

(C) $$f_n(y) = g(y, n) \frac{d^n q(y, n)}{d[h(y)]^n}, \quad g(y, n) \not\equiv f_n(y),$$

where n is a positive integer?

<u>Transformations</u>:

(17) $$e^y {}_1F_1(b - a; b; -y) = {}_1F_1(a; b; y),$$

(18) $$e^y {}_0F_1(a + \tfrac{1}{2}; \tfrac{1}{4} y^2) = {}_1F_1(a; 2a; 2y),$$

(19) $$\Gamma(a + 1) y^{-\frac{a}{2}} J_a(2\sqrt{y}) = {}_0F_1(a + 1; -y),$$

(20) $$(1 - y)^{-a} F(a, c - b; c; \tfrac{y}{y-1}) = F(a, b; c; y).$$

To formulate an inquiry into a general class of transformations which includes all the formulas (17) through (20) is rather difficult, because it is hard to find a clear distinction between them and an entirely arbitrary functional transformation. Tentatively we suggest

Question 4. Do there exist functions $h(n, y)$ and $q(y)$ such that

$$\text{(D)} \qquad h(n, y)f_n(q(y)) = \sum_{n=0}^{\infty} a_n y^n,$$

where the a_n can be exhibited explicitly? The word "explicitly" leaves considerable vagueness in this question.

Contour Integrals:

$$\text{(21)} \qquad J_a(y) = \frac{(\frac{1}{2}y)^a}{2\pi i} \int_{-\infty}^{(0+)} w^{-a-1} \exp(w - \frac{y^2}{4w}) dw,$$

$$\text{(22)} \quad J_a(y) = \frac{\Gamma(\frac{1}{2} - a)(\frac{1}{2}y)^a}{2\pi i \ \Gamma(\frac{1}{2})} \int_A^{(1+,-1-)} (w^2-1)^{a-\frac{1}{2}} \cos wy \, dw,$$

$$\text{(23)} \qquad P_a(y) = \frac{1}{2^a 2\pi i} \int_A^{(1+,-1-)} (w^2-1)^a (w-y)^{-a-1} dw,$$

$$\text{(24)} \qquad P_a(y) = \frac{1}{\pi} \int_0^{\pi} [y + \sqrt{y^2 - 1} \cos w]^{a+1} dw,$$

$$\text{(25)} \qquad L_n^{(a)}(y) = \frac{\Gamma(a+n+1)}{n! \ 2\pi i} \int_{-\infty}^{(0+)} (1 - \frac{y}{w})^n w^{-a-1} e^w \, dw.$$

Supposing $f_n(x)$ belongs to our class of functions, it is natural to ask:

Question 5. Can we find a contour integral for $f_n(x)$?

Definite Integrals:

$$\text{(26)} \qquad \int_0^{\infty} t^{a-b} J_a(ct) dt = \frac{2^{a-b}\Gamma(\frac{a}{2} + \frac{1}{2})}{c^{a-b+1}\Gamma(b - \frac{a}{2} + \frac{1}{2})},$$

(27)

$$\int_0^\infty e^{-at} J_c(bt) t^{d-1} dt = \frac{b^c \Gamma(c+d)}{2^c a^{c+d} \Gamma(c+1)} F\left(\frac{c+d}{2}, \frac{c+d+1}{2}; c+1; -\frac{b^2}{c^2}\right),$$

(28)

$$\int_0^\infty e^{-b^2 t^2} t^{d-1} J_c(at) dt = \frac{\Gamma(\frac{c+d}{2}) a^c}{\Gamma(c+1) 2^{c+1} b^{c+d}} {}_1F_1\left(\frac{c+d}{2}; c+1; -\frac{a^2}{4b^2}\right),$$

(29) $$\int_0^\infty e^{-bt} t^a L_m^{(a)}(t) dt = \frac{\Gamma(a+m+1)(b-1)^m}{m!\; b^{a+m+1}} .$$

Question 6. Can the integral

(E) $$\int_0^\infty e^{-bt^c} t^a f_n(t) dt$$

be evaluated in general? What other integrals involving $f_n(y)$ is it natural to set up?

Integro-difference Relations:

(30) $$y^{\frac{b+a}{2}} J_{b+a}(2\sqrt{y}) = \frac{1}{\Gamma(b)} \int_0^y w^{\frac{1}{2}a} (y-w)^{b-1} J_a(2\sqrt{w}) dw,$$

(31) $$y^{-\frac{a-b-1}{2}} J_{a-b-1}(2\sqrt{y}) = \frac{1}{\Gamma(b+1)} \int_0^\infty t^b (y+t)^{-\frac{1}{2}a} J_a(2\sqrt{y+t}) dt,$$

(32) $$L_c^{(a+b)}(y) = \frac{\Gamma(a+b+c+1)}{\Gamma(b)\Gamma(a+c+1)} \int_0^1 t^a (1-t)^{b-1} L_c^{(a)}(ty) dt .$$

Question 7. For a given function $f_n(y)$ can suitable functions $g(w, a, n)$ and $h(w, y)$ and numbers b and c be found such that

(F) $$f_{n+a}(y) = \int_b^c g(w, a, n)\, f_n(h(w, y)) dy \;?$$

Miscellaneous Relations:

$$(33)\qquad e^{-y}y^{\frac{1}{2}a}L_b^{(a)}(y) = \frac{1}{\Gamma(b+1)}\int_0^\infty e^{-t}t^{b+\frac{a}{2}}J_a(2\sqrt{ty})dt$$

$$(34)\qquad P_a(y) = \frac{1}{\sqrt{\pi}\,\Gamma(a+1)}\int_{-\infty}^\infty e^{-t^2}t^aH_a(ty)dt,$$

$$(35)\qquad P_a^b(y) = \frac{1}{\Gamma(a-b+1)}\int_0^\infty e^{-ty}J_b(t\sqrt{1-y^2})t^a dt.$$

Question 8. Is there a general way to find a relation expressing one individual member of our class of functions in terms of a second individual member?

We shall try to answer Questions 1 to 8. In other words, The aim of this essay is to provide a general theory which motivates, discovers, and coordinates such seemingly unconnected relations among familiar special functions as the formulas (1) through (35).

I emphasize the words "motivate" and "discover." It is no great task to construct *ad hoc* rigorous proofs and to find the range of validity of any of the formulas we have just listed, once one sees them set up; their derivations form typical exercises for the student in English text books. No text book, however, suggests as a problem "Find a formula which gives Laguerre polynomials in terms of Bessel functions," because the student would have no idea where to start unless he had the advantage of having seen such a formula before. The discoverers of these formulas have used their intuition combined often with brilliant artifices and have discovered them singly, giving us slight indication how to find others like them. In this essay consequently, I aim to give rational methods of discovery; if limitation of space, inattention, or ignorance has not always permitted me to satisfy the reader's standards of rigor, still I believe that it is possible to supply the missing details and to correct any erroneous ones.

I emphasize also the word "coordinate." It is my aim to bring order into a part of the collection of known relations concerning special functions by showing that they are simple special cases of about a dozen general formulas and by adding to their number some of the missing analogues which do not seem to have been discovered thus far.

Finally I emphasize the words "familiar special functions." This essay does not consider an equation for its own sake and then seek functions satisfying it which may serve as examples, but rather it observes and investigates a hitherto largely neglected property common to the transcendents occurring most frequently in mathematical physics and shows that it is this very property which entails as necessary consequences such relations as the formulas (1) to (35).

Looking at a typical collection of 35 formal relations involving familiar special functions, we have listed eight questions which come to mind immediately, and we have seen that in order to give them any real meaning we must refer them to the theory of some suitable class of functions. Our first problem, then, is to find this class. We wish to include Bessel, Legendre, Laguerre, and Hermite functions, so let us examine the major formal properties these functions have in common: (1) They satisfy ordinary linear differential equations of the second order; (2) they satisfy ordinary linear difference equations of the second order; (3) with suitable weight functions they form complete sets of orthogonal functions over suitable intervals; (4) they satisfy linear differential-difference relations. The first three of these properties after very long and thorough investigations by numerous excellent mathematicians have yielded but slight clews to the discovery of such relations as the formulas (1) to (35). Our confidence in the essential beauty and perfection of classical analysis would be shaken if in fact these formulas were, so to speak, random effusions of the Divine Mathematician, disjoint and chaotic, so we are driven to conclude that they are consequences of some of the differential recurrence relations

$$\frac{\partial}{\partial y} J_a(y) = \frac{a}{y} J_a(y) - J_{a+1}(y), \tag{36}$$

$$\frac{\partial}{\partial y} J_a(y) = J_{a-1}(y) - \frac{a}{y} J_a(y), \tag{37}$$

$$(1-y^2)\frac{\partial}{\partial y} P_a^b(y) = (a-b+1)(yP_a^b(y) - P_{a+1}^b(y)), \tag{38}$$

$$(1-y^2)\frac{\partial}{\partial y} P_a^b(y) = (a+b)P_{a-1}^b(y) - ayP_a^b(y), \tag{39}$$

$$(1-y^2)\frac{\partial}{\partial y} P_a^b(y) = -\sqrt{1-y^2}\, P_a^{b+1}(y) - by\, P_a^b(y), \tag{40}$$

$$(1-y^2)\frac{\partial}{\partial y} P_a^b(y) = (a-b+1)(a+b)\sqrt{1-y^2}\, P_a^{b-1}(y) + byP_a^b(y), \tag{41}$$

$$\frac{\partial}{\partial y} L_a^{(b)}(y) = L_a^{(b)}(y) - L_a^{(b+1)}(y), \tag{42}$$

$$y\frac{\partial}{\partial y} L_a^{(b)}(y) = (a+b)L_a^{(b-1)}(y) - bL_a^{(b)}(y), \tag{43}$$

$$y\frac{\partial}{\partial y} L_a^{(b)}(y) = aL_a^{(b)}(y) - (a+b)L_{a-1}^{(b)}(y), \tag{44}$$

$$y\frac{\partial}{\partial y} L_a^{(b)}(y) = (a+1)L_{a+1}^{(b)}(y) - (a+b+1-y)L_a^{(b)}(y), \tag{45}$$

$$\frac{\partial}{\partial y} H_a(y) = 2yH_a(y) - H_{a+1}(y), \tag{46}$$

$$\frac{\partial}{\partial y} H_a(y) = 2aH_{a-1}(y). \tag{47}$$

Accordingly we shall study functional equations of this type.

§2. THE METHOD AND GENERAL RESULTS OF THIS ESSAY. THE F-EQUATION

We are going to study functions $f(y, \alpha)$ satisfying a functional equation of the type

$$(1) \quad \frac{\partial}{\partial y} f(y, \alpha) = A(y, \alpha)\, f(y, \alpha) + B(y, \alpha)\, f(y, \alpha+1),$$

where $A(y, \alpha)$ and $B(y, \alpha)$ are given functions, and $B(y, \alpha)$ does not vanish identically. We shall show that by a proper transformation, which we shall exhibit explicitly, this equation may be reduced to the form

$$(2) \quad \frac{\partial}{\partial y} g(y, \alpha) = C(y, \alpha)\, g(y, \alpha+1),$$

and that when $C(y, \alpha) = Y(y)\, A(\alpha)$ this second equation by means of a second transformation, which also we shall exhibit explicitly, is in turn reducible to the form

$$(3) \quad \frac{\partial}{\partial z} F(z, \alpha) = F(z, \alpha+1).$$

The present essay centers around this functional equation. To emphasize it, and to facilitate the discourse, we shall henceforth call the equation (3) "the F-equation."

Proper transforms of many different familiar special functions satisfy <u>the same equation</u>: the F-equation. Even though being a solution of the F-equation is a common property of so many different functions, being a solution of the F-equation is often a more convenient defining property for a particular function than is being a solution of a suitable ordinary differential equation. For example, as we shall see, we may define $J_a(x)$ by saying $e^{i\alpha\pi} z^{-\alpha/2} J_\alpha(2\sqrt{z})$ is the unique solution of the F-equation which reduces to $e^{i\alpha\pi}/\Gamma(\alpha+1)$ when $z = 0$, $R\alpha > -1$, and conclude immediately that $z^{-\alpha/2} J_\alpha(2\sqrt{z})$ is an integral function of z with the power series expansion

$$z^{-\alpha/2} J_\alpha(2\sqrt{z}) = \sum_{n=0}^{\infty} \frac{(-)^n z^n}{n!\Gamma(\alpha+n+1)};$$

this much, I say, we shall know at a glance.

We shall show that the formulas (1.2), (1.3), (1.4), and (1.5) are all special cases of a simple generating expansion, common to all analytic solutions of the F-equation, and that all possess a generalization analogous to the formula (1.10). We shall be able to find other general expansions which yield as special cases the formulas (1.6), (1.8), and (1.9), which give us some information about the formula (1.1), and which suggest an approach to the formula (1.7). Hence we shall adequately answer Questions 1 and 2 for certain values of the a_n (§14).

We shall show that the formulas (1.11), (1.13), (1.15), and (1.16) are all special cases of a single n-th derivative formula. While we shall discover the formulas (1.12) and (1.14) incidentally as special cases of contour integrals which we derive in another connection, we shall show that they are not analogues of the other n-th derivative formulas in the list (1.11) to (1.16). Hence we shall answer Question 3 (§17).

We shall show that the transformations (1.17), (1.18), (1.19), and (1.20) are all very easy special cases of a single power series expansion, and hence we shall completely answer Question 4 (§11).

We shall give a method of discovery which leads in a straightforward fashion to the contour integrals (1.21) to (1.25). Hence we shall answer Question 5 (§13).

We shall show that the formula (1.26) is a special case of a general infinite integral involving a class of solutions of the F-equation, and that the formulas (1.27), (1.28), (1.29), (1.31), (1.33), and (1.35) are all special cases of a second integral formula. We shall mention other definite integrals whose integrands contain solutions of the F-equation, and sketch a method for building up integral formulas in increasing complexity. Hence we shall answer Question 6 adequately (§15).

We shall show that the formulas (1.30), (1.31), and (1.32) are all special cases of the same integro-difference relation. Hence we shall answer Question 7 (§18).

We shall give a straightforward general method of attacking such problems as "Find a formula giving the Laguerre function $L_b^{(a)}(y)$ in terms of the Bessel function $J_c(y)$" which will lead automatically to the formula (1.33), and will serve equally well to find an inverse for it. The method will with equal ease discover the formula (1.34) and an inverse for it. It does not seem to apply easily to the formula (1.35), which, however we may discover very simply by means of another general formula satisfied by a class of solutions of the F-equation. Hence we may go far to answer Question 8 (§16).

We shall make some headway in coordinating formal relations satisfied by familiar special functions by showing that a great many of them are included as special cases of about a dozen general theorems concerning solutions of the F-equation. For example, we shall see that the proper analogue for the Bessel functions of the Legendre function formula (1.2) is not the formula (1.1) but the formula (1.5).

While answering these questions we shall find other topics of interest to be investigated, and in the end we shall have a fair insight into some of the formal properties of familiar special functions.

It is not my aim to produce a long list of new relations satisfied by various special functions, but rather to render trivial the discovery and proof of a large class of these formulas. As illustrations I have applied the results of this investigation to deduce sometimes a formula I had already known, and sometimes one previously unknown to me which I had discovered (quite possibly rediscovered) by the methods here presented.

I hope that I shall be forgiven for listing a rather large number of examples. The methods we discuss in this essay are so easy and elementary that without inductive evidence to the contrary I fear the reader might doubt they could be good for anything.

Chapter II

REDUCTION TO THE F-EQUATION

§3. REDUCTION TO THE F-EQUATION

Suppose the function $f(y, \alpha)$ satisfies the functional equation

$$(1) \qquad \frac{\partial}{\partial y} f(y, \alpha) = A(y, \alpha)\, f(y, \alpha) + B(y, \alpha)\, f(y, \alpha+1).$$

Let the function $g(y, \alpha)$ be defined by the transformation

$$(2) \qquad g(y, \alpha) \equiv \exp\left\{- \int_{y_0}^{y} A(v, \alpha)dv\right\} f(y, \alpha).$$

Then it is easy to verify that $g(y, \alpha)$ satisfies the equation

$$(3) \qquad \frac{\partial}{\partial y} g(y, \alpha) = B(y, \alpha) \exp\left\{\int_{y_0}^{y} \Delta_{\alpha} A(v, \alpha)dv\right\} g(y, \alpha+1).$$

Hence all equations of the type (1) may be reduced to the form

$$(4) \qquad \frac{\partial}{\partial y} g(y, \alpha) = C(y, \alpha)\, g(y, \alpha+1).$$

In the case of nearly every special function that I know to satisfy an equation of type (4), the coefficient $C(y, \alpha)$ is factorable:

$$(5) \qquad C(y, \alpha) = Y(y)\, A(\alpha).$$

The restriction imposed upon the coefficients of the original equation (1) by the factorability condition (5) will be stated in Theorem [7.1]. Suppose, then, that the function $g(y, \alpha)$ satisfies the equation

$$(6) \qquad \frac{\partial}{\partial y} g(y, \alpha) = Y(y)\, A(\alpha)\, g(y, \alpha+1).$$

Let us define z and $F(z, \alpha)$ by the transformations

$$(7) \qquad z \equiv \int_{y_1}^{y} Y(v)dv,$$

$$(8) \qquad F(z, \alpha) \equiv \exp \{ \mathop{\mathbf{S}}_{\alpha_0}^{\alpha} \log A(v)\, \Delta v \}\, g(y, \alpha).$$

Then

$$\frac{\partial}{\partial z} F(z,\alpha) = \exp \{ \mathop{\mathbf{S}}_{\alpha_0}^{\alpha} \log A(v) \Delta v \} Y(y) A(\alpha) \frac{dy}{dz} g(y,\alpha+1);$$

$$= \exp \{ \mathop{\mathbf{S}}_{\alpha_0}^{\alpha+1} \log A(v) \Delta v - \mathbf{\Delta}_{\alpha} \mathop{\mathbf{S}}_{\alpha_0}^{\alpha} \log A(v) \Delta v \} A(\alpha) g(y,\alpha+1),$$

$$= \exp \{ \mathop{\mathbf{S}}_{\alpha_0}^{\alpha+1} \log A(v) \Delta v \} g(y,\alpha+1),$$

$$= F(z, \alpha+1).$$

Hence the equation

$$(9) \qquad \frac{\partial}{\partial z} F(z, \alpha) = F(z, \alpha+1),$$

the F-equation, is as general as the equation (6). That the factorability condition (5) is not only sufficient but also necessary for the reduction of equation (1) to the F-equation we shall see in Theorem [7.1].

In operational notation the F-equation has the form

$$(10) \qquad (\mathbf{D}_z - \mathbf{E}_\alpha)\, F(z, \alpha) = 0.$$

If $G(z, \alpha) \equiv e^{-z} F(z, \alpha)$, where $F(z, \alpha)$ satisfies the F-equation, then

$$(D_z - \Delta_\alpha)\, G(z, \alpha) = 0; \tag{11}$$

this form is convenient for some considerations.

In Appendix III are listed the only transforms of familiar functions which I know to satisfy an equation of type (4) which is <u>not</u> reducible to the F-equation.

§4. NOTATION

We now formalize the conventions of notations which we have observed up to the present and shall continue to observe.

<u>Functions</u>. Capital F always stands for a solution of F-equation. For other classes of functions, no particular notation is adopted.

<u>Complex Variables and Parameters</u>. Latin u, v, w, y, a, b, c, d, with or without subscripts, represent general complex variables or parameters. The letters z and α represent the variables in the F-equation, assumed complex unless otherwise specified. Thus $F(z, \alpha)$ always means a solution of the F-equation.

<u>Real Variables</u>. Latin x, t, T, k, h represent real variables or parameters.

<u>Positive Integers</u>. Latin i, j, m, n, p, q, r represent positive integers.

<u>Special Functions</u>. Our notation for special functions is roughly that of Whittaker and Watson 1. Precise definitions are given in Appendix I.

§5. SOLUTIONS OF THE F-EQUATION DERIVED FROM FAMILIAR FUNCTIONS

By starting with the particular functional equations of the type (3.1) which are satisfied by familiar special functions and performing upon them the reductions of §3, we may accumulate particular solutions of the F-equation. Below is a partial list. In each case, z and α are the variables of the F-equation, and any other parameters, such as, $a, b, c,$ are understood to be independent of z and α. In the column "Name and Reference" we give the name of the function involved and a reference to the particular functional equation or to a preceding solution used as a starting point in deriving the particular solution.

	Expression	Name and Reference
1.	e^z	Exponential function.
2.	$\sin(z - \frac{1}{2}\alpha\pi)$	Sine function.
3.	$e^{i\alpha\pi}\Gamma(\alpha)z^{-\alpha}$	Power function.
4.	$\frac{\Gamma(\alpha+a_1)\Gamma(\alpha+a_2)\ldots\Gamma(\alpha+a_p)}{\Gamma(\alpha+b_1)\Gamma(\alpha+b_2)\ldots\Gamma(\alpha+b_q)}\, {}_pF_q(\alpha+a_1, \alpha+a_2, \ldots, \alpha+a_p; \alpha+b_1, \alpha+b_2, \ldots, \alpha+b_q; z)$	Generalized hypergeometric function. (C.8).
5.	$\frac{\Gamma(\alpha+b)\Gamma(\alpha-c)}{\Gamma(\alpha)}(1-z)^{-\alpha-b}F(\alpha+b, c; \alpha; \frac{z}{z-1})$	Hypergeometric function. Solution 4 and (1.20), or (C.13) only.

6. $\frac{\Gamma(\alpha-b)\Gamma(\alpha-c)}{\Gamma(\alpha)}(1-z)^{b+c-\alpha}F(b, c; \alpha; z)$

Hypergeometric function. Solution 4 and (C.16), or (C.12) only.

7. $e^{i\alpha\pi}\Gamma(\alpha+1)z^{-\alpha-1}F(b,c;-\alpha;z)$ Hypergeometric function. (C.10).

8. $e^{i\alpha\pi}\Gamma(\alpha+1)z^{-\alpha-1}F(b,-\alpha-c;-\alpha;\frac{z}{z-1})$

Hypergeometric function. Solution 7 and (1.20).

9. $e^{i\alpha\pi}\Gamma(\alpha+1)z^{-\alpha-1}(1-z)^{-\alpha-b-c}F(-\alpha-b,-\alpha-c;-\alpha;z)$

Hypergeometric function. Solution 7 and (C.16), or (C.15) only.

10. $e^{i\alpha\pi}\Gamma(\alpha)z^{-\alpha}F(\alpha,c;b;\frac{1}{z})$ Hypergeometric function. (C.9).

11. $\Gamma(\alpha+c)z^{-c}(1-z)^{-\alpha+b-c}F(-\alpha,b;c;\frac{1}{z})$

Hypergeometric function. Solution 10 and (1.20) or (C.16), or (C.11) only.

12. $e^{i\alpha\pi}\Gamma(\alpha+1)z^{-\alpha-1}(1-z)^{-b+\alpha+1}F(-\alpha-c,b;-\alpha;z)$

Hypergeometric function. (C.14).

13. $e^{i\alpha\pi}\Gamma(\alpha+1)z^{-\alpha-1}(1-z)^{\alpha+1}F(c,b;-\alpha;\frac{z}{z-1})$

Hypergeometric function. Solution 12 and (1.20).

14. $e^{i\alpha\pi}\Gamma(\alpha+1)z^{-\alpha-1}(1-z)^{b+2\alpha+\gamma+1}F(-\alpha-c,-\alpha-b;-\alpha;\frac{z}{z-1})$

Hypergeometric function. Solution 12 and (1.20).

15. $\Gamma(\alpha-b+1)(z^2+1)^{-\frac{\alpha+1}{2}}P_\alpha^b(-\frac{z}{\sqrt{z^2+1}})$

Associated Legendre Function. (1.38).

16. $\Gamma(\alpha-b+1)(z^2+1)^{-\frac{\alpha+1}{2}}Q_\alpha^b(-\frac{z}{\sqrt{z^2+1}})$

Associated Legendre Function. (1.38).

17. $\Gamma(\alpha-b)(z^2+1)^{-\frac{\alpha}{2}}P_{-\alpha}^b(-\frac{z}{\sqrt{z^2+1}})$

Associated Legendre Function. (1.39), or Solution 15 and (D.24).

18. $\Gamma(\alpha-b)(z^2+1)^{-\frac{\alpha}{2}}Q_{-\alpha}^b(-\frac{z}{\sqrt{z^2+1}})$

Associated Legendre Function. (1.39).

19. $(1-z^2)^{-\frac{\alpha}{2}}P_b^\alpha(z)$

Associated Legendre Function. (1.40).

20. $(1-z^2)^{-\frac{\alpha}{2}}Q_b^\alpha(z)$

Associated Legendre Function. (1.40).

21. $\Gamma(\alpha+b+1)\Gamma(b-\alpha)(1-z^2)^{-\frac{\alpha}{2}}P_b^{-\alpha}(z)$

Associated Legendre Function. (1.41).

22. $\Gamma(\alpha+b+1)\Gamma(b-\alpha)(1-z^2)^{-\frac{\alpha}{2}}Q_b^{-\alpha}(z)$

Associated Legendre Function. (1.41).

23. $e^{i\alpha\pi}\Gamma(\alpha)z^{-\alpha-\frac{1}{2}b}C_b^{\alpha}(\frac{1}{\sqrt{z}})$ Gegenbauer function. (D.10).

24. $\Gamma(\alpha+1)(z^2+1)^{-\frac{\alpha+1}{2}}(\frac{\sqrt{z^2+1}-z}{\sqrt{z^2+1}+z})^{-\frac{1}{2}b}P_{\alpha}^{(b,-b)}(-\frac{z}{\sqrt{z^2+1}})$

Jacobi function. (D.3).

25. $\Gamma(\alpha)(z^2+1)^{-\frac{\alpha}{2}}(\frac{\sqrt{z^2+1}-z}{\sqrt{z^2+1}+z})^{-\frac{1}{2}b}P_{-\alpha}^{(b,-b)}(-\frac{z}{\sqrt{z^2+1}})$

Jacobi function. (D.2).

26. $e^{i\alpha\pi}e^{-z}L_b^{(\alpha)}(z)$ Laguerre function. (1.42).

27. $\dfrac{z^{-\alpha}L_b^{(-\alpha)}(z)}{\Gamma(b-\alpha+1)}$ Laguerre function. (1.43).

28. $\Gamma(\alpha+1)(-z)^{-\alpha-1-b}e^{-\frac{1}{z}}L_{\alpha}^{(b)}(\frac{1}{z})$ Laguerre function. (1.45).

29. $\Gamma(\alpha-b)(-z)^{-\alpha}L_{-\alpha}^{(b)}(\frac{1}{z})$ Laguerre function. (1.44).

30.	$\dfrac{e^{-2\sqrt{z}}}{\Gamma(\alpha + \frac{1}{2})} \, {}_1F_1(\alpha;\ 2\alpha;\ 4\sqrt{z})$	Confluent hypergeometric function. (E.19).
31.	$e^{i\alpha\pi} z^{-\alpha} \gamma(\alpha,\ z)$	Incomplete gamma function. (E.18).
32.	$\Gamma(\alpha + \frac{1}{2} + c)\,\Gamma(\alpha + \frac{1}{2} - c)(-z)^{-\alpha} e^{\frac{1}{2}z} W_{-\alpha, c}(\frac{1}{z})$	Whittaker function. (E.7).
33.	$e^{-z^2} H_{\alpha}(-z)$	Hermite function. (1.46).
34.	$\Gamma(\alpha) H_{-\alpha}(-\frac{1}{2} z)$	Hermite function. (1.47).
35.	$\psi_{\alpha}(b,\ z)$	Poisson-Charlier function. (E.16).
36.	$e^{z} \psi_b(-\alpha,\ z)$	Poisson-Charlier function. (E.15).
37.	$e^{i\alpha\pi} z^{-\alpha/2} J_{\alpha}(2\sqrt{z})$	Bessel function. (1.36).
38.	$e^{i\alpha\pi} z^{-\alpha/2} Y_{\alpha}(2\sqrt{z})$	Bessel function. (1.36).
39.	$e^{i\alpha\pi} z^{-\alpha/2} H_{\alpha}^{(1)}(2\sqrt{z})$	Hankel function. (1.36).
40.	$e^{i\alpha\pi} z^{-\alpha/2} H_{\alpha}^{(2)}(2\sqrt{z})$	Hankel function. (1.36).
41.	$z^{-\alpha/2} J_{-\alpha}(2\sqrt{z})$	Bessel function. (1.37).
42.	$z^{-\alpha/2} Y_{-\alpha}(2\sqrt{z})$	Bessel function. (1.37).
43.	$z^{-\alpha/2} H_{-\alpha}^{(1)}(2\sqrt{z})$	Hankel function. (1.37).
44.	$z^{-\alpha/2} H_{-\alpha}^{(2)}(2\sqrt{z})$	Hankel function. (1.37).
45.	$\phi(e^{z},\ -\alpha)$	Spence's transcendent. (G.1).

46. $\dfrac{\phi^{(b)}_{-\alpha}(z)}{(-\alpha)!}$ — ϕ-polynomial (α is a negative integer). (G.3).

47. $\psi_{\alpha}(z)$ — Polygamma function (α is a positive integer). (G.6).

48. $e^{i\alpha\pi}\,\Gamma(\alpha)\,\zeta(\alpha, z)$ — Generalized zeta function. (G.8).

We have used a variety of notations as a matter of convenience. All the three variable transcendents occurring in solutions Nos. 15 to 25, for example, could be replaced by suitable hypergeometric functions $F(a,b;c;z)$, and all the two variable transcendents occurring in solutions Nos. 26 to 29 and 31 to 44 by suitable linear combinations or limits of such combinations of confluent hypergeometric functions ${}_1F_1(a;b;y)$, but we have used the most concise form rather than the most basic in each instance.

The methods of the following sections will enable us to deduce many other solutions of the F-equation from those in this list. We should notice here, however, that if $F(z, \alpha)$ satisfies the F-equation, so does $F_1(z, \alpha)$ as given by the transformation

$$(1) \qquad F_1(z,\alpha) = \pi(\alpha)\, b^{\alpha}\, F(bz + a,\ \alpha + c),$$

where a, b, and c are any quantities independent of z and α, and $\pi(\alpha + 1) = \pi(\alpha)$. Thus, for example, by the choice $b = i$ from the solution No. 15 we may deduce another which is sometimes more convenient,

$$\Gamma(\alpha - b + 1)(z^2 - 1)^{-\frac{\alpha+1}{2}}\, P^{b}_{\alpha}\left(-\frac{z}{\sqrt{z^2-1}}\right).$$

In Corollary [7.3] we shall prove that the simple transformation (1) is the only change of variable of the type

$$F_1(z, \alpha) \equiv M(z, \alpha)\, F(h(z), \alpha+c)$$

which from a given solution $F(z, \alpha)$ of the F-equation will deduce a second solution $F_1(z, \alpha)$.

§6. FUNCTIONS WHICH SATISFY ALSO AN ORDINARY DIFFERENTIAL OR DIFFERENCE EQUATION OF THE SECOND ORDER

The recurrence relations satisfied by the special functions of potential theory are but special cases of relations of contiguity among various solutions of the generalized Lamé equation, those of the type that may be reduced to the F-equation arising as the result of some confluence of singularities. While it would be possible to deduce a certain body of our results by studying properties of solutions of the generalized Lamé equation, not only should we unnecessarily encumber ourselves but also we should lose generality. The class of solutions of the F-equation is not included in the class of solutions of any class of simple differential equations. Bell has shown that his exponential polynomials $\xi_n(y, a; m)$ satisfy no linear differential equation in y of order independent of m and y, the coefficients of which are polynomials in m, y, a, and n, provided $m > 2$. [Bell 1, p. 262.] It is easy to verify that $\exp(xz^r)\, \xi_\alpha(x, z; r)$ is a solution of the F-equation. The theory of solutions of the F-equation is therefore completely separate from the theory of solutions of ordinary differential equations.

We shall find it useful however, to take note of one property of functions which satisfy not only the F-equation but also a linear differential equation of the second order in z, or a linear difference equation of the second order in α. Suppose $f(y, \alpha)$ is a solution of the the functional equation (3.1). If

(1) $$g(y, \alpha) \equiv f(y, -\alpha)$$

then

$$(2) \quad \frac{\partial}{\partial y} g(y, \alpha) = A(y, -\alpha)\, f(y, -\alpha) + B(y, -\alpha)\, f(y, -\alpha+1),$$

$$= A(y, -\alpha)\, g(y, \alpha) + B(y, -\alpha)\, g(y, \alpha-1).$$

Now $g(y, \alpha)$ will satisfy a functional equation of the type (3.1) if and only if there are functions $D(y, \alpha)$ and $E(y, \alpha)$ such that

$$(3) \quad \frac{\partial}{\partial y} g(y, \alpha) = D(y, \alpha)\, g(y, \alpha) + E(y, \alpha)\, g(y, \alpha+1).$$

Hence by comparing equations (2) and (3), we see that the necessary and sufficient condition becomes

$$(4) \quad \begin{aligned} &E(y, \alpha)\, g(y, \alpha+1) + [D(y, \alpha) - A(y, -\alpha)]\, g(y, \alpha) \\ &\qquad - B(y, -\alpha)\, g(y, \alpha-1) = 0 , \end{aligned}$$

or, equivalently,

$$(5) \quad \begin{aligned} &B(y, \alpha)\, f(y, \alpha+1) + [A(y, \alpha) - D(y, -\alpha)]\, f(y, \alpha) \\ &\qquad - E(y, -\alpha)\, f(y, \alpha-1) = 0 . \end{aligned}$$

Therefore a necessary and sufficient condition that not only $f(y, \alpha)$ but also $f(y, -\alpha)$ satisfy a functional equation of type (3.1) is that $f(y, \alpha)$ satisfy also a homogeneous linear difference equation of the second order in α. Obviously an equivalent condition is that $f(y, \alpha)$ satisfy also a homogeneous linear differential equation of the second order in y. Most of the familiar functions making up the solutions listed in §5 satisfy this condition.

The solutions of the F-equation which are derived from the transcendents of potential theory are thus members of a special subclass of solutions. From those which like the Laguerre functions, satisfy two ordinary difference equations, one in each of two parameters, there may be derived four independent solutions of the F-equation. The two-variable transcendents of number theory from which solutions of the F-equation may be derived do not appear to satisfy any difference equations, and consequently furnish us with but one solution of the F-equation apiece. While our subsequent results will refer to the F-equation alone and will make no use of the fact that a subclass of solutions of the greatest interest satisfy a difference equation as well, it is partly the consequent multiplicity of solutions derived from a single function which makes our results more easily applicable to the transcendents of potential theory than to the two-variable transcendents of number theory.

Although we shall study only solutions of the F-equation and shall make no use of difference or differential equations, we shall find it useful in the construction of particular solutions of the F-equation to have an explicit formula exhibiting a solution of the F-equation derived from $F(z, -\alpha)$ when $F(z, \alpha)$ is a solution of both the F-equation and of a linear difference equation of second order.

THEOREM [6.1]. Suppose $F_1(z, \alpha)$ satisfies both the F-equation and the difference equation

$$(6) \quad F_1(z,\alpha+1) = [b(z)\alpha+c(z)]F_1(z,\alpha) + h(z)k(\alpha)F_1(z,\alpha-1).$$

Let w as a function of z be defined by the integral

$$(7) \quad z = \int_{w_0}^{w} h(v) \exp \{- \int_{w_1}^{v} b(u)du\}dv.$$

Then the function $F_2(z, \alpha)$ given by the definition

$$(8) \quad F_2(z,\alpha)=\exp\{\mathop{S}_{\alpha_0}^{\alpha} \log k(-v)\Delta v-\int_{w_1}^{w} [-b(v)\alpha+c(v)]dv\}F_1(w,-\alpha)$$

is also a solution of the F-equation. Here α_0, w_0, and w_1 are any convenient constants.

Proof: If

$$M(w, \alpha) \equiv \exp\{\mathop{S}_{\alpha_0}^{\alpha} \log k(-v)\Delta v - \int_{w_1}^{w} [-b(v)\alpha + c(v)]dv\},$$

then

$$F_2(z, \alpha) = M(w, \alpha)\, F_1(w, -\alpha),$$

and

$$\frac{\partial}{\partial z}F_2(z,\alpha) = M(w,\alpha)[\{b(w)\alpha - c(w)\}F_1(w,-\alpha) + F_1(w,-\alpha+1)]\frac{dw}{dz},$$

since $F_1(w, \alpha)$ satisfies the F-equation. Since $F_1(w, \alpha)$ satisfies also the difference equation (6), we may see that

$$\frac{\partial}{\partial z} F_2(z, \alpha) = M(w, \alpha) \frac{dw}{dz} h(w)\, k(-\alpha)\, F_1(w, -\alpha-1).$$

Now

$$\frac{dw}{dz} = \frac{1}{\frac{dz}{dw}},$$

$$= \frac{\exp\{\int_{w_1}^{w} b(u)du\}}{h(w)},$$

by the formula (7). Hence

$$\frac{\partial}{\partial z} F_2(z,\alpha) = M(w,\alpha)k(-\alpha) \exp\{\int_{w_1}^{w} b(u)du\}F_1(w,-\alpha-1)$$

$$= \exp\{\mathop{S}_{\alpha_0}^{\alpha} \log k(-v)\Delta v + \log k(-\alpha)$$

$$- \int_{w_1}^{w} (-b(u)\alpha+c(u)]du + \int_{w_1}^{w} b(u)du\}F_1(w,-\alpha-1),$$

$$= \exp\{ \mathop{S}_{\alpha_0}^{\alpha+1} \log k(-v)\Delta v - \mathop{\Delta}_{\alpha} \mathop{S}_{\alpha_0}^{\alpha} \log k(-v)\Delta v + \log k(-\alpha)$$

$$- \int_{w_1}^{w} [-b(u)(\alpha+1)+c(u)]du\}F_1(w,-\alpha-1),$$

$$= M(w,\alpha+1)F_1(w,-\alpha-1),$$

$$= F_2(z,\alpha+1). \qquad \text{Q.E.D.}$$

The equation (6) is perhaps not the most general difference equation which a function need satisfy in order for us to be able to exhibit a formula like the transformation (8), but it is easy to treat and it covers all the special cases with which I am familiar.

COROLLARY [6.2]. In the formula (6), if

$$(9) \qquad b(z) = \frac{h'(z)}{h(z)},$$

then the formula (8) of Theorem [6.1] becomes simply

$$(10) \quad F_2(z,\alpha) = [h(z)]^{\alpha} \exp\{\mathop{S}_{\alpha_0}^{\alpha} \log k(-v)\Delta v - \int_{w_1}^{w} c(u)du\}F_1(z,-\alpha).$$

All the special functions which I know to satisfy the conditions of Theorem [6.1] satisfy also the condition (9).

The formula (10) was used to find some of the entries in the table of §5.

§7. THE UNIQUENESS OF THE REDUCTIONS OF §3

It is natural to ask whether a given function $f(y, \alpha)$ which satisfies an equation of the type (3.1) may be transformed in more than one way into a function satisfying an equation of the type (3.4). Let us restrict ourselves, as in §3, to changes of variable of the type

$$(1) \qquad g(y, \alpha) \equiv M(y, \alpha)\, f(h(y), \alpha).$$

Then

$$\frac{\partial}{\partial y} g(y,\alpha) = [\frac{\partial}{\partial y} M(y,\alpha) + M(y,\alpha)h'(y)A(h(y),\alpha)]\, f(h(y),\alpha)$$

$$+ B(h(y),\alpha)M(y,\alpha)h'(y)f(h(y),\alpha+1).$$

In order for this equation to be of the type (3.4) it is necessary and sufficient that

$$\frac{1}{M(y, \alpha)} \frac{\partial}{\partial y} M(y, \alpha) = -h'(y)A(h(y), \alpha),$$

or

$$M(y, \alpha) = H(\alpha) \exp\{-\int_{h(y_0)}^{h(y)} A(v, \alpha)dv\} .$$

It follows that if $h(y)$ and $H(\alpha)$ are quite arbitrary functions, then the transformation

$$(2) \qquad g(y, \alpha) \equiv H(\alpha) \exp\{-\int_{h(y_0)}^{h(y)} A(v, \alpha)dv\}\, f(h(y), \alpha)$$

will define a function $g(y, \alpha)$ satisfying an equation of the type (3.4). In fact

$$(3) \qquad \frac{\partial}{\partial y} g(y,\alpha)$$

$$= \frac{H(\alpha)}{H(\alpha+1)} \exp\{\int_{h(y_0)}^{h(y)} \Delta_\alpha A(v,\alpha)dv\}\, B(h(y),\alpha)h'(y)g(y,\alpha+1) .$$

There are, then, infinitely many ways of reducing an equation of type (3.1) to one of type (3.4), with infinitely many different resulting coefficients $C(y, \alpha)$.

It is next natural to ask whether the factorability condition (3.5) was a necessary one for the reduction of the equation (3.1) to the F-equation, and what restriction this condition imposes upon the coefficients $A(y, \alpha)$ and $B(y, \alpha)$ in the equation (3.1). Let us answer the second question first. In view of equation (3), the factorability condition implies that

$$Y(y)A(\alpha) = \frac{H(\alpha)}{H(\alpha+1)} \exp\left\{ \int_{h(y_0)}^{h(y)} \underset{\alpha}{\Delta} A(v,\alpha)dv \right\} B(h(y),\alpha)h'(y).$$

In this relation let us replace y by $h^{-1}(y)$, where $h^{-1}(y)$ is a function inverse to $h(y)$. Then if $Y_1(y) \equiv Y(h^{-1}(y))/h'(y)$, the previous relation becomes simply

$$Y_1(y)A(\alpha) = \frac{H(\alpha)}{H(\alpha+1)} \exp\left\{ \int_{y_0}^{y} \underset{\alpha}{\Delta} A(v,\alpha)dv \right\} B(y,\alpha).$$

Hence

$$Y_1'(y)A(\alpha)$$

$$= \frac{H(\alpha)}{H(\alpha+1)} \exp\left\{ \int_{y_0}^{y} \underset{\alpha}{\Delta} A(v,\alpha)dv \right\} [B(y,\alpha) \underset{\alpha}{\Delta} A(y,\alpha) + \frac{\partial}{\partial y} B(y,\alpha)],$$

or

$$Y_1'(y)A(\alpha) = Y_1(y)A(\alpha)[\underset{\alpha}{\Delta} A(y,\alpha) + \frac{\partial}{\partial y} \log B(y,\alpha)].$$

Hence a necessary condition for the factorability of $C(y, \alpha)$ is that there exist a function $L(y)$ such that

$$(4) \qquad \underset{\alpha}{\Delta} A(y, \alpha) + \underset{y}{D} \log B(y, \alpha) = L(y),$$

or, equivalently, that there exist functions $s(y)$ and $t(y)$ such that

(5) $$A(y, \alpha) = \alpha s(y) + \pi(\alpha)\, t(y) - \underset{\alpha_0}{\overset{\alpha}{S}} \underset{y}{D} \log B(y, v)\Delta v$$

where $\pi(\alpha)$ is a periodic function of period 1. It is easy to show that the condition (5) is also a sufficient one for the factorability of $C(y, \alpha)$. The conditions (4) and (5) are conditions on the coefficients in the original equation (3.1).

The first question is even more straightforward to answer. If an equation of the type (3.1) is reducible to the F-equation by a transformation of type (1), then the F-equation may be transformed back to the original equation by another transformation of type (1). The most general equation reducible to the F-equation is then the most general equation into which the F-equation may be transformed. Suppose $F(z, \alpha)$ satisfies the F-equation, and let $f(y, \alpha)$ be defined by the transformation (1):

$$f(y, \alpha) \equiv M(y, \alpha)\, F(h(y), \alpha).$$

Then

$$\frac{\partial}{\partial y} f(y,\alpha) = F(h(y),\alpha) \frac{\partial}{\partial y} M(y,\alpha) + M(y,\alpha)h'(y)F(h(y),\alpha+1),$$

$$= \left[\frac{1}{M(y,\alpha)} \frac{\partial}{\partial y} M(y,\alpha)\right] f(y,\alpha) + \frac{h'(y)M(y,\alpha)}{M(y,\alpha+1)} f(y,\alpha+1).$$

Hence

$$A(y, \alpha) = \frac{1}{M(y, \alpha)} \frac{\partial}{\partial y} M(y, \alpha),$$

$$B(y, \alpha) = h'(y) \frac{M(y, \alpha)}{M(y, \alpha+1)}.$$

From the form of these two equations it is easy to verify that the condition (4) is satisfied. Hence we have proved the following theorem:

THEOREM [7.1]. For an equation of type (3.1) to be reducible to the F-equation by a transformation of type (1), it is necessary and sufficient that the coefficients satisfy the condition (5), which is equivalent to the factorability condition (3.5) for the coefficient $C(y, \alpha)$ in any equation of type (3.4) to which the original equation may be reduced.

We are now in a position to prove that if an equation is reducible to the F-equation, it is so reducible in an essentially unique fashion.

THEOREM [7.2]. Suppose a function $f(y, \alpha)$ satisfies an equation of type (3.1) which is reducible to the F-equation. Then if $F_1(z, \alpha)$ and $F_2(z, \alpha)$ are two transforms of type (1) of the same function $f(y, \alpha)$ which both satisfy the F-equation, there exist a periodic function $\pi(\alpha)$ of period 1 and constants b and c such that

$$F_2(z, \alpha) = \pi(\alpha) b^{\alpha} F_1(bz + c, \alpha).$$

Proof: Suppose $f(y, \alpha)$ satisfies an equation of type (3.1). If this equation is reducible to the F-equation, by Theorem [7.1] its coefficient $A(y, \alpha)$ may be eliminated by the condition (5). Suppose there is a solution $F(z, \alpha)$ of the F-equation such that

$$F(z, \alpha) = M(z, \alpha)\, f(h(z), \alpha) .$$

Then

$$\frac{\partial}{\partial z} F(z,\alpha) = f(h(z),\alpha) \frac{\partial}{\partial z} M(z,\alpha) + M(z,\alpha)h'(z) \cdot$$

$$\{[\alpha s(h(z)) + \pi_0(\alpha)t(h(z)) - \mathop{S}_{\alpha_0}^{\alpha} D_{h(z)} \log B(h(z),v)\Delta v]f(h(z),\alpha)$$

$$+ B(h(z),\alpha)\, f(h(z),\alpha+1)\} ,$$

where we have used the relation (5). The functions $s(z)$ and $t(z)$ are not at our disposal, being given by the coefficients $A(z, \alpha)$ and $B(z, \alpha)$ of the original equation of type (3.1). But if $F(z, \alpha)$ satisfies the F-equation,

$$\frac{\partial}{\partial z} F(z, \alpha) = M(z, \alpha+1)\, f(h(z), \alpha+1).$$

By substituting this relation on the left in the preceding equation and equating the coefficients of $f(h(z), \alpha)$ and $f(h(z), \alpha+1)$ on each side we conclude that

(A) $D_z \log M(z,\alpha) =$

$$-[\alpha s(h(z)) + \pi_0(\alpha)t(h(z)) - \mathop{S}_{\alpha_0}^{\alpha} D_{h(z)} \log B(h(z),w)\Delta w]h'(z)$$

and

(B) $$M(z, \alpha+1) = B(h(z), \alpha)\, h'(z)\, M(z, \alpha).$$

From the relation (A) we see that

$M(z, \alpha)$

$$= T(\alpha) \exp\{-\int_{h(z_0)}^{h(z)} [\alpha s(v) + \pi_0(\alpha)t(v) - \mathop{S}_{\alpha_0}^{\alpha} D_v \log B(v,w)\Delta w]dv$$

where $T(\alpha)$ is an arbitrary function, while from this result and the relation (B) we see that

$$T(\alpha+1)\exp\{-\int_{h(z_0)}^{h(z)}[s(v)-D_v\log B(v,\alpha)]dv\}$$

$$= T(\alpha)\, B(h(z),\,\alpha)\, h'(z)\ ,$$

or

$$\frac{T(\alpha+1)}{T(\alpha)} = h'(z)\exp\int_{h(z_0)}^{h(z)} s(v)dv\ .$$

Hence there exists a constant b such that

$$\text{(C)}\qquad h'(z)\exp\int_{h(z_0)}^{h(z)} s(v)dv = b$$

and

$$\text{(D)}\qquad \frac{T(\alpha+1)}{T(\alpha)} = b.$$

From the difference equation (D) it follows that

$$T(\alpha) = \pi(\alpha)\, b^{\alpha},$$

where $\pi(\alpha)$ is a periodic function of period 1. If in the relation (C) we make the substitution $y = h(z)$ and then integrate, we see that

$$\text{(E)}\qquad bz = \int_{y_0}^{y}[\exp\int_{y_0}^{w} s(v)dv]dw.$$

Hence our most general transformation of type (1) which will reduce $f(y,\alpha)$ to a solution of the F-equation is

$$\text{(F)}\qquad F(z,\alpha) = \pi(\alpha)b^{\alpha}\exp\{-\int_{y_0}^{y}[\alpha s(v)+\pi_0(\alpha)t(v)$$

$$-\mathop{S}_{\alpha_0}^{\alpha} D_v\log B(v,w)\Delta w]dv\ f(y,\alpha)\ ,$$

where y as a function of z is given by the formula (E).

Now suppose we have two transforms $F_1(z, \alpha)$ and $F_2(z, \alpha)$ of the <u>same</u> function $f(y, \alpha)$ which both satisfy the F-equation. Then there exist $\pi_1(\alpha)$, b_1, and y_{01} such that $F_1(z, \alpha)$ can be deduced from $f(y, \alpha)$ by the formula (F). (Choice of the constants z_0 and α_0 gives us no added generality.) There exist also $\pi_2(\alpha)$, b_2, and y_{02} for $F_2(z, \alpha)$. Let us regard the relation (E) as giving z in terms of y, and say that z_1 corresponds to $F_1(z, \alpha)$ and z_2 to $F_2(z, \alpha)$. Then

$$b_1 \frac{dz_1}{dy} = b_2 \frac{dz_2}{dy} .$$

Hence

$$z_1 = \frac{b_2}{b_1} z_2 + c,$$

where c is a constant. Let us say $b \equiv b_2/b_1$ and $\pi(\alpha) \equiv \pi_2(\alpha)/\pi_1(\alpha)$. By the relation (F),

$$F_2(z_2, \alpha) = \pi_2(\alpha) b_2^{\alpha} \exp\{-\int_{y_0}^{y} [\alpha s(v) + \pi_0(\alpha) t(v)$$

$$- \underset{\alpha_0}{\overset{\alpha}{S}} \underset{v}{D} \log B(v, w) \Delta w] dv\} f(y, \alpha),$$

$$= \frac{\pi_2(\alpha)}{\pi_1(\alpha)} \left(\frac{b_2}{b_1}\right)^{\alpha} \pi_1(\alpha) b_1^{\alpha} \exp\{-\int_{y_0}^{y} [\alpha s(v) + \pi_0(\alpha) t(v)$$

$$- \underset{\alpha_0}{\overset{\alpha}{S}} \underset{v}{D} \log B(v, w) \Delta w] dv\} f(y, \alpha),$$

$$= \pi(\alpha) b^{\alpha} F_1(z_1, \alpha),$$

$$= \pi(\alpha) b^{\alpha} F_1(bz_1 + c, \alpha). \qquad \text{Q.E.D.}$$

COROLLARY [7.3]. If $F(z, \alpha)$ is a solution of the F-equation, any function deduced from it by a transformation of type (1) is a solution of the F-equation if and only if it is of the form

$$\pi(\alpha) b^{\alpha} F(bz + c, \alpha).$$

This corollary shows that the list in §5 of solutions of the F-equation involving familiar functions cannot be enlarged except trivially by changes of variable of the type (1).

§8. AN INTEGRAL EQUATION WHICH IN SOME CASES IS EQUIVALENT TO THE EQUATION (3.4).

We have seen that when $C(y, \alpha)$ is factorable the equation (3.4) is reducible to the F-equation. We now notice in passing that under certain restrictions on the function $C(y, \alpha)$ the equation (3.4) is equivalent to a linear integral equation. Since these restrictions are not satisfied if $C(y, \alpha) \equiv 1$, the following theorem does not belong to the theory of the F-equation.

THEOREM [8.1]. Suppose $g(y, \alpha)$ satisfies equation (3.4), and suppose there exist sectionally continuous functions $G(w, b)$ and $K(w, b)$, of exponential order in both variables, such that

$$g(y, \alpha) = \int_0^\infty dt\, e^{-\alpha t} \int_0^\infty dx\, e^{-yx}\, G(x, t), \tag{1}$$

$$C(y, \alpha) = \int_0^\infty dt\, e^{-\alpha t} \int_0^\infty dx\, e^{-yx}\, K(x, t); \tag{2}$$

then $G(w, b)$ is a solution of the linear integral equation

$$-wG(w, b) = \int_0^b dt \int_0^w dv\, e^{-t}\, K(w-x, b-t)\, G(x, t). \tag{3}$$

Conversely, if $K(w, b)$ and $G(w, b)$ are sectionally continuous functions of exponential order in both variables which are related through the integral equation (3), then $g(y, \alpha)$ as defined by equation (1) is a solution of a functional equation of the type (3.4) in which the coefficient $C(y, \alpha)$ is defined by the formula (2).

Proof: I. By hypothesis,

$$\int_0^\infty dt e^{-\alpha t} \int_0^\infty dx(-x)e^{-yx}G(x,t)$$

$$= [\int_0^\infty dt e^{-\alpha t} \int_0^\infty dx e^{-yx}K(x,t)][\int_0^\infty dt e^{-(\alpha+1)t} \int_0^\infty dx e^{-yx}G(x,t)],$$

$$= \int_0^\infty dt e^{-\alpha t} \int_0^t du[\int_0^\infty dx e^{-yx}K(x,t-u)][\int_0^\infty dx e^{-yx-u}G(x,u)].$$

Hence

$$- \int_0^\infty dx v e^{-yx}G(x,t)$$

$$= \int_0^t du[\int_0^\infty dx e^{-yx}K(x,t-u)][\int_0^\infty dx e^{-yx-u}G(x,u)],$$

$$= \int_0^t du \int_0^\infty dx e^{-yx} \int_0^x dw K(x-w,t-u)e^{-u}G(x,u),$$

$$= \int_0^\infty dx e^{-yx} \int_0^t du \int_0^x dw K(x-w,t-u)e^{-u}G(x,u).$$

Hence

$$-xG(x,t) = \int_0^t du \int_0^x dw K(x-w,t-u)e^{-u}G(x,u). \qquad \text{Q.E.D.}$$

II. By hypothesis,

$$-wG(w,b) = \int_0^b dv \int_0^w dte^{-x}K(w-t,b-x)G(t,x).$$

Then

$$\int_0^\infty dwe^{-yw}G(w,b)$$

$$= - \int_0^\infty \frac{dw}{w} e^{-yw} \int_0^b dx \int_0^w dte^{-x}K(w-t,b-x)G(t,x),$$

$$= \int_0^b dxe^{-x} \int_0^\infty dt \int_t^\infty dw[- \frac{e^{-yw}}{w} K(w-t,b-x)]G(t,x),$$

$$= \int_0^b dxe^{-x} \int_0^\infty dtG(t,x) \int_0^\infty du \frac{-e^{-y(t+u)}}{t+u} K(u,b-x).$$

Then

$$\frac{\partial}{\partial y} [\int_0^\infty dwe^{-yw}G(w,b)]$$

$$= \int_0^b dxe^{-x} \int_0^\infty dte^{-yt}G(t,x) \int_0^\infty due^{-yu}K(u,b-x).$$

Then

$$\frac{\partial}{\partial y} [\int_0^\infty dbe^{-\alpha b} \int_0^\infty dwe^{-yw}G(w,b)]$$

$$= \int_0^\infty dbe^{-\alpha b} \int_0^b dxe^{-x} \int_0^\infty dte^{-yt}G(t,x) \int_0^\infty due^{-yu}K(u,b-x);$$

$$= \int_0^\infty dbe^{-\alpha b} \int_0^b dxe^{-x}[\int_0^\infty dte^{-yt}G(t,x)][\int_0^\infty due^{-yu}K(u,b-x)],$$

$$= [\int_0^\infty dbe^{-b-\alpha b} \int_0^\infty dte^{-yt}G(t,b)][\int_0^\infty dbe^{-\alpha b} \int_0^\infty due^{-yu}K(u,b)].$$

Q.E.D.

Example [8.1]. Consider the equation

$$(4) \qquad \frac{\partial}{\partial y} F(y, \alpha) = \frac{e^{-\sqrt{\alpha y}}}{\sqrt{\alpha y}} F(y, \alpha+1).$$

Now

$$\frac{e^{-\sqrt{\alpha y}}}{\sqrt{\alpha y}} = \int_0^\infty dt\ e^{-\alpha t} \frac{e^{-\frac{y}{4t}}}{\sqrt{\pi t}\ y},$$

$$= \int_0^\infty dt\ e^{-\alpha t} \int_0^\infty dx\ e^{-yx} K(x, t),$$

where [Churchill 1, pp. 298, 299, nos. 84, 61]

$$K(x, t) = \begin{cases} 0 \quad , & 0 < x < \frac{1}{4t} ; \\ \frac{1}{\sqrt{\pi t}} \quad , & x > \frac{1}{4t} . \end{cases}$$

Hence a linear integral equation equivalent to the functional equation (4) is

$$(5) \qquad -wG(w, b) = \frac{e^b}{\sqrt{\pi}} \int_0^b dx \frac{e^{-x}}{\sqrt{x}} \int_{\frac{1}{4v}}^{w} dt\ G(w-t, b-x) .$$

Chapter III

EXISTENCE AND UNIQUENESS THEOREMS

§9. THE EXISTENCE AND UNIQUENESS OF SOLUTIONS OF A MORE GENERAL FUNCTIONAL EQUATION

We now state fundamental existence and uniqueness theorems for a functional equation much more general than the equation (3.1), namely

$$(1)\ \frac{\partial}{\partial x} \vec{f}(x,\alpha) = \vec{G}(\vec{f}(x,\alpha+n),\ \vec{f}(x,\alpha+n-1),\ \ldots,\ \vec{f}(x,\alpha),x,\alpha),$$

where the notation $\vec{a}$ stands for the p-vector $(a_1,\ a_2,\ \ldots,\ a_p)$, where $\vec{G}(\vec{t}^{(n)},\ \vec{t}^{(n-1)},\ \ldots,\ \vec{t}^{(o)},\ x,\alpha)$ is a given real vector function of its $n + 3$ arguments when x and α are real variables and $\vec{t}^{(o)}\vec{t}^{(1)}\ldots\vec{t}^{(n)}$ are real p-vectors. This equation reduces to the equation (3.1) when $p = 1$, $n = 1$, and

$$(2)\quad G_1(\vec{t}^{(1)},\ \vec{t}^{(o)},\ x,\ \alpha) = A(x,\ \alpha)t_1^{(o)} + B(x,\ \alpha)t_1^{(1)}.$$

It does not include the typical differential-difference equations of the dynamics of constrained systems of particles, which are of the form [Bateman 3, p. 494]

$$(3)\ M\frac{\partial^2}{\partial x^2} f(x,\alpha) + 2K\frac{\partial}{\partial x} f(x,\alpha) + Sf(x,\alpha)$$

$$= (\alpha a+a+b)\ \underset{\alpha}{\Delta}\ f(x,\alpha) - (\alpha a+c)\ \underset{\alpha}{\Delta}\ f(x,\alpha-1),$$

where M, K, S, a, b, c are constants, except in the very special case when $M = 0$, $a = 0$, $c = 0$, which coincides with the special case of the equation (3.1) when the coefficients $A(x,\ \alpha)$ and $B(x,\ \alpha)$ are constants.

We shall confine ourselves to the case when x and α are real variables. The corresponding theorems for complex variables should not be essentially different. We shall use geometric language so that the conditions of the theorem may readily be visualized.

We shall employ the absolute value metric:

$$|\vec{T} - \vec{t}| = \sum_{i=1}^{p} |T_i - t_i| . \tag{4}$$

We state existence and uniqueness conditions for the condition

$$\vec{f}(x_0, \alpha) = \vec{\phi}(\alpha), \tag{5}$$

where $\vec{\phi}(\alpha)$ is defined either discretely when $\alpha = \alpha_0 + m$, $m = 0, 1, 2, \ldots,$ or continuously when $\alpha \geqq \alpha_0$.

THEOREM [9.1]. Let α and x be real variables. Let $\vec{\phi}(\alpha)$ be defined when $\alpha = \alpha_0 + m$, $m = 0, 1, 2, \ldots,$ where α_0 is some fixed number. In the real $p(n+1) + 2$ space of points $(\vec{t}^{(n)}, \vec{t}^{(n-1)}, \ldots, \vec{t}^{(o)}, x, \alpha)$ let U_m be the region of the plane $\alpha = \alpha_0 + m$ defined by the inequalities

$$|\vec{t}^{(j)} - \vec{\phi}(\alpha_0 + m + j)| \leqq b, \; j = 0, 1, 2, \ldots, n,$$

and $|x - x_0| \leqq a$, where x_0, a, and b are given fixed real numbers. Let us write

$$U = \bigcup_{m=0}^{\infty} U_m .$$

In U, suppose that

(a) $\vec{G}(\vec{t}^{(n)}, \vec{t}^{(n-1)}, \ldots, \vec{t}^{(o)}, x, \alpha)$ is a given real single-valued function, such that, when $\vec{t}^{(o)}(x), \vec{t}^{(1)}(x), \ldots, \vec{t}^{(n)}(x)$ are absolutely integrable functions of x, then $\vec{G}(\vec{t}^{(n)}(x), \vec{t}^{(n-1)}(x), \ldots, \vec{t}^{(o)}(x), x, \alpha)$ is also an absolutely integrable function of x.

(b) $|\vec{G}(\vec{t}^{(n)}, \vec{t}^{(n-1)}, \ldots, \vec{t}^{(o)}, x, \alpha)| < M$,

where M is a fixed constant.

Let h be the lesser of a and b/M; call H the cylinder in which $|x-x_o| \leqq h$ and write $V \equiv U \cap H$. If the two points $(\vec{T}^{(n)}, \vec{T}^{(n-1)}, \ldots, \vec{T}^{(o)}, x, \alpha)$ and $(\vec{t}^{(n)}, \vec{t}^{(n-1)}, \ldots, \vec{t}^{(o)}, x, \alpha)$ lie in V, suppose further that $\vec{G}$ satisfies the Lipschitz condition

(c) $$|\vec{G}(\vec{T}^{(n)}, \vec{T}^{(n-1)}, \ldots, \vec{T}^{(o)}, x, \alpha) - \vec{G}(\vec{t}^{(n)}, \vec{t}^{(n-1)}, \ldots, \vec{t}^{(o)}, x, \alpha)| \leqq K \sum_{m=0}^{n} |\vec{T}^{(m)} - \vec{t}^{(m)}| ,$$

where K is a real constant independent of x and α. Let W_m be the line segment of the $x\alpha$ plane defined by the inequality $|x - x_o| \leqq h$, $\alpha = \alpha_o + m$, where m is a nonnegative integer, and let W be the union of the W_m:

$$W = \bigcup_{m=0}^{\infty} W_m .$$

Then when $(x, \alpha) \in W$ there exists a unique single-valued function $\vec{f}(x, \alpha)$ which satisfies the equation (1) and the boundary condition (5).

The regions U, V, and W are shown in Figure 1 for the case when $n = 1$, $p = 1$, and $G_1(t_1^{(1)}, t_1^{(o)}, x, \alpha)$ is independent of $t_1^{(o)}$. The geometrical intuition of the reader, with this assistance, will readily conceive these regions in general, although their formal definition has been lengthy.

In the succeeding sketch of the proof we shall often use without explicit mention the obvious fact that if $(x, \alpha) \in W$, then $(x, \alpha + m) \in W$, where m is any positive integer.

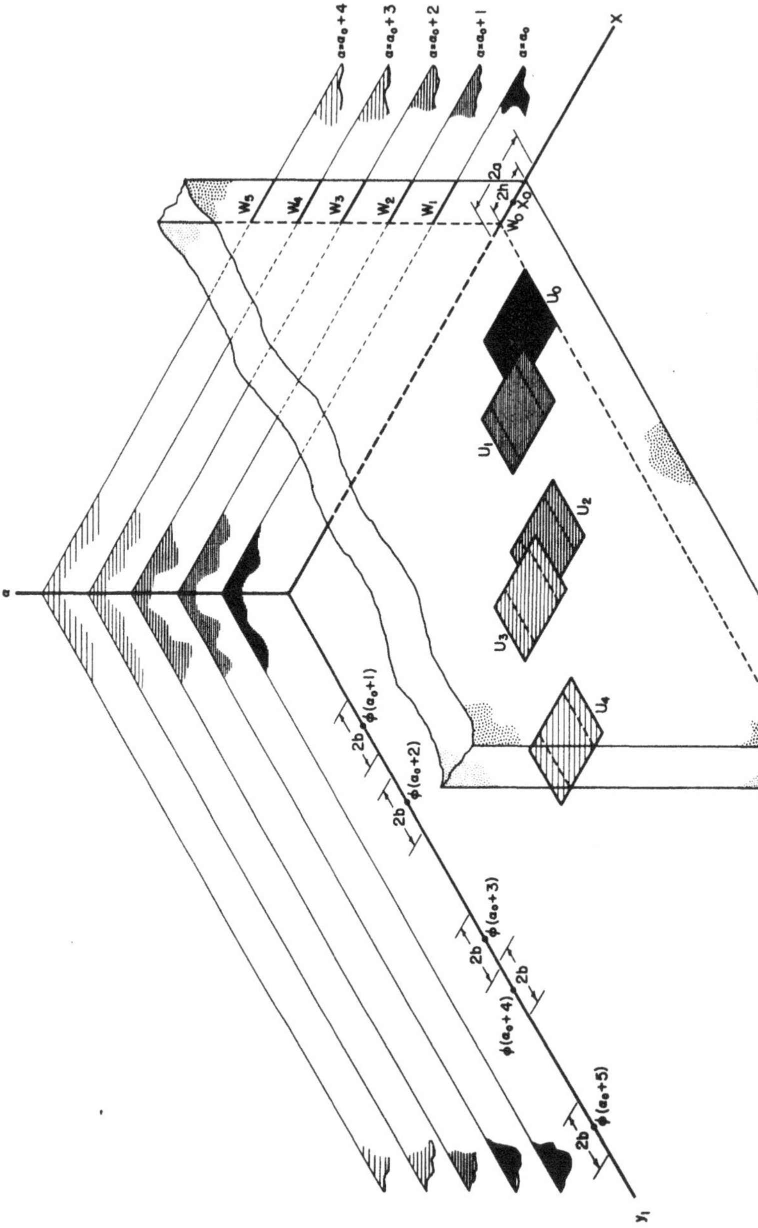
a
X
y_1
$a=a_0+4$
$a=a_0+3$
$a=a_0+2$
$a=a_0+1$
$a=a_0$
W_5
W_4
W_3
W_2
W_1
W_0
x_0
2a
2h
U_0
U_1
U_2
U_3
U_4
H
$\phi(a_0+1)$
$\phi(a_0+2)$
$\phi(a_0+3)$
$\phi(a_0+4)$
$\phi(a_0+5)$
2b

FIGURE I

Outline of Proof: Let $\vec{f}_0(x, \alpha)$ be an arbitrary real vector function, absolutely integrable with respect to x whenever $(x, \alpha) \in W$, such that $|\vec{f}_0(x, \alpha) - \vec{\phi}(\alpha)| \leqq b$ and $|\vec{f}_0(x, \alpha) - \vec{\phi}(\alpha)| < M|x - x_0|$ whenever $(x, \alpha) \in W$. Certainly such functions exist, since a possible choice for $\vec{f}_0(x, \alpha)$ is $\vec{\phi}(\alpha)$.

Formally let us define $\vec{f}_{r+1}(x, \alpha)$ by the relation

$$\text{(A)} \quad \vec{f}_{r+1}(x,\alpha)$$

$$= \vec{\phi}(\alpha) + \int_{x_0}^{x} \vec{G}(\vec{f}_r(t,\alpha+n),\vec{f}_r(t,\alpha+n-1),\ldots,\vec{f}_r(t,\alpha),t,\alpha)dt \; .$$

We must first show that the integral in the formula (A) exists; in view of assumption (a), it will be sufficient to show that the point $(\vec{f}_r(x, \alpha+n), \vec{f}_r(x, \alpha+n-1), \ldots, \vec{f}_r(x, \alpha), x, \alpha)$ lies in U, whatever the value of r, provided $(x, \alpha) \in W$. We shall in fact show that it lies in V. For this purpose it will suffice to show that

$$|\vec{f}_r(x,\alpha_0+j+m) - \vec{\phi}(\alpha_0+j+m)| < b, \; m=0,1,\ldots; \; j=0,1,\ldots,n \; ,$$

when $|x - x_0| \leqq h$, or, more simply,

$$\text{(B)} \quad |\vec{f}_r(x,\alpha) - \vec{\phi}(x,\alpha)| \leqq b \quad \text{if} \quad (x,\alpha) \in W, \; r = 0, 1, \ldots .$$

We shall prove this inequality by induction. The function $\vec{f}_0(x, \alpha)$ by hypothesis is chosen so that the inequality (B) holds for it. Suppose now

$$|\vec{f}_{r-1}(x, \alpha) - \vec{\phi}(\alpha)| \leqq b \quad \text{if} \quad (x, \alpha) \in W,$$

where r is fixed. Then if $(x, \alpha) \in W$ the integral (A) defining $\vec{f}_{r+1}(x, \alpha)$ exists, and

$$|\vec{f}_r(x,\alpha) - \vec{\phi}(\alpha)|$$

$$\leqq \int_{x_0}^{x} |\vec{G}(\vec{f}_{r-1}(t,\alpha+n),\vec{f}_{r-1}(t,\alpha+n-1),\ldots,\vec{f}_{r-1}(t,\alpha),t,\alpha)|dt.$$

Then by assumption (b), since the argument system in the integrand is by hypothesis in U,

$$|\vec{f}_r(x, \alpha) - \vec{\Phi}(\alpha)| \leqq M|x - x_0| ,$$
$$\leqq Mh,$$
$$\leqq b,$$

whenever $(x, \alpha) \in W$. The induction is now complete. It is obvious that the $\vec{f}_r(x, \alpha)$ are continuous functions of x.

A consequence of the inequality (B) is that

$$\text{(C)} \quad (\vec{f}_r(x, \alpha+n), \vec{f}_r(x, \alpha+n-1), \ldots, \vec{f}_r(x, \alpha), x, \alpha) \in V$$

whenever $(x, \alpha) \in W$.

The next step in the proof is with the aid of the result (C) to prove by induction that

$$\text{(D)} \quad |\vec{f}_{r+1}(x, \alpha) - \vec{f}_r(x, \alpha)| \leqq \frac{2MK^r(n+1)^r}{(r+1)!} |x - x_0|^{r+1}$$

whenever $(x, \alpha) \in W$. This step and the remainder of the existence and uniqueness proof we omit, because it is a straightforward generalization of the classical successive approximation proof for the existence and uniqueness of solutions of an ordinary differential equation of first order [Ince 1, pp. 62-66], of which it is a direct generalization.

Like the sufficient conditions used in the statement of the classical existence theorem for an ordinary differential equation of first order, the sufficient conditions we have used in the statement of Theorem [9.1] are not the weakest we might have used, but they are far weaker than we shall find necessary for the treatment of interesting special cases.

Bruwier has stated and proved an existence and uniqueness theorem equivalent to our Theorem [9.1] in the special case when $n = 1$ and $p = 1$, and has asserted that its extension to our more general case offers no difficulty [Bruwier 1, pp. 43-48]. He does not distinguish between the case when the boundary condition is satisfied only when $\alpha = \alpha_0 + m$, which we have just considered, and the case when the boundary condition is satisfied when $\alpha \geqq \alpha_0$, which we are about to consider.

THEOREM [9.2]. Let U_y be the region of the plane $\alpha = \alpha_0 + y$ defined by the inequalities

$$|\vec{t}^{(m)} - \vec{\Phi}(\alpha_0+y+m)| \leqq b, \; m = 1, 2, \ldots, n,$$

and $|x - x_0| \leqq a$, where x_0, a, b, and y are real constants. Let U_{ex} be the union of the U_y:

$$U_{ex} \equiv \bigcup_{y>0} U_y.$$

Let W_{ex} be the semi-infinite strip of the $x\alpha$ plane where $|x - x_0| \leqq h$ and $\alpha \geqq \alpha_0$. Let us use the abbreviation $\bar{t} \equiv (\vec{t}^{(n)}, \vec{t}^{(n-1)}, \ldots, \vec{t}^{(o)}, \alpha)$; then $\vec{G}(\vec{t}^{(n)}, \vec{t}^{(n-1)}, \ldots, \vec{t}^{(o)}, x, \alpha) = \vec{G}(\bar{t}, x)$. In the statement of the previous theorem replace U by U_{ex} and W by W_{ex}. Then if

(d) $\vec{\Phi}(\alpha)$ is a continuous function of α when $\alpha \geqq \alpha_0$,

$(b)_{ex}$ $\vec{G}(\bar{t},x)$ is a continuous function of $\bar{t}$ when $(\bar{t},x) \in U_{ex}$,

the solution of the system (1) satisfying the boundary condition (5) whose existence and uniqueness were proved in the previous theorem when $(x, \alpha) \in W$ exists and is unique whenever $(x, \alpha) \in W_{ex}$, and is a continuous function of α.

Note: In Figure 1 the region U_{ex} may be visualized as the interior and boundary of a surface passing through the boundaries of all the rectangles shown and a continuum of rectangles passing continuously from one to the next.

Sketch of Proof: What we have done is essentially to pass a surface through the regions U_m and W_m of Theorem [9.1], and to extend our conditions on the function $\vec{G}$ and the variables so as to apply within these surfaces. Hence a unique solution exists, and it remains only to show it is a continuous function of α.

I. We shall prove by induction that when $(x, \alpha) \in W_{ex}$, $\vec{f}_r(x, \alpha)$ is a continuous function of α. Now our limit function $\vec{f}(x, \alpha)$, being unique, is independent of the choice of (permitted) initial functions $\vec{f}_o(x, \alpha)$; hence we may suppose without loss of generality that we choose a function $\vec{f}_o(x, \alpha)$ which is a continuous function of α, such a choice being possible since $\vec{\Phi}(\alpha)$ itself is now by hypothesis continuous. Suppose now $\vec{f}_r(x, \alpha)$, for some fixed r, is a continuous function of α when $(x, \alpha) \in W_{ex}$. Then the function $\bar{Y}(x, \alpha)$, where,

$$\bar{Y}(x, \alpha) \equiv (\vec{f}_r(x, \alpha+n), \vec{f}_r(x, \alpha+n-1), \ldots, \vec{f}_r(x, \alpha), \alpha)$$

is also a continuous function of α, and hence by the assumption $(b)_{ex}$ so is $\vec{G}(\bar{Y}(x, \alpha), x)$. It follows from the definition (A) that $\vec{f}_{r+1}(x, \alpha)$ is also a continuous function of α when $(x, \alpha) \in W_{ex}$.

II. It is evident from the inequality (D) that $\vec{f}_r(x, \alpha) \longrightarrow \vec{f}(x, \alpha)$ uniformly in α; hence $\vec{f}(x, \alpha)$ is a continuous function of α.

The real concern of this essay is not the equation (1), but equations of type (3.4) and the F-equation, for which it is consequently worthwhile to write out the much simpler forms the existence and uniqueness theorems assume.

COROLLARY [9.3]. (REAL EXISTENCE THEOREM FOR THE EQUATION (3.4).) Suppose

(i) $C(x, \alpha)$ is defined at least when $|x - x_0| \leqq a$, $\alpha = \alpha_0 + m$, $m = 0, 1, 2, \ldots$;

(ii) $C(x, \alpha)$ is absolutely integrable with respect to x;

(iii) $|C(x, \alpha)| < M_1$;

(iv) $|\phi(\alpha_0 + m)| < M_2$.

Let h be the lesser of a and $1/M_1$. Then there exists a unique function $g(x, \alpha)$ such that

(v) $\frac{\partial}{\partial x} g(x, \alpha) = C(x, \alpha)\, g(x, \alpha+1)$;

(vi) $g(x_0, \alpha) = \phi(\alpha)$;

(vii) $g(x, \alpha)$ is a continuous function of x, whenever $|x - x_0| < h$ and $\alpha = \alpha_0 + m$, $m = 0, 1, 2, \ldots$.

Proof: Choose any number b, and form the region U as in Theorem [9.1]. When $|\vec{t}_1^{(1)} - \vec{\phi}_1(\alpha_0+1+m)| \leqq b$, then $|\vec{t}_1^{(1)}| \leqq b + M_2$. Now

$$\vec{G}_1(\vec{t}_1^{(n)}, \vec{t}_1^{(n-1)}, \ldots, \vec{t}_1^{(0)}, x, \alpha) = C(x, \alpha)\vec{t}_1^{(1)}.$$

Hence under our hypotheses, in U

$$|\vec{G}(\vec{t}^{(n)}, \vec{t}^{(n-1)}, \ldots, \vec{t}^{(0)}, x, \alpha)| < M_1(b + M_2).$$

Then if h is defined as in Theorem [9.1],

$$h = \min(a, \frac{b}{M_1(b+M_2)}) .$$

But b is quite arbitrary, so we may let it become indefinitely large. Hence $h = \min(a, 1/M_1)$. The Lipschitz condition is automatically satisfied. Hence all the hypotheses of Theorem [9.1] are true in this special case. Q.E.D.

COROLLARY [9.4]. (STRONG UNIQUENESS THEOREM FOR THE EQUATION (3.4).) When $C(x, \alpha)$ satisfies the conditions of Corollary [9.3], there cannot exist more than one function $g(x, \alpha)$ satisfying the equation (v) and the condition $g(x_0, \alpha) = \phi(\alpha)$, whether or not $\phi(\alpha)$ is bounded.

Proof: By Theorem [9.1], when $|\phi(\alpha_0+m)|$ is bounded as $m \longrightarrow \infty$, the solution is unique. Suppose now all conditions are removed from $\phi(\alpha)$ except that it be defined when $\alpha = \alpha_0 + m$, $m = 0, 1, 2, \ldots.$ Suppose there are two solutions $g_1(x, \alpha)$ and $g_2(x, \alpha)$, both of them satisfying both the equation (v) and the boundary condition (vi) when $|x - x_0| \leqq a$ and $\alpha = \alpha_0 + m$, $m = 0, 1, 2, \ldots.$ Then if $g(x, \alpha) \equiv g_1(x, \alpha) - g_2(x, \alpha)$, $g(x, \alpha)$ satisfies the equation (v) in the same region, and fulfills the boundary condition $g(x_0, \alpha) = 0$. Now 0 is a bounded function of α, so Theorem [9.1] is again applicable, and since 0 is plainly a solution of the equation (v) satisfying the boundary condition $g(x_0, \alpha) = 0$, it follows that $g(x, \alpha) \equiv 0$. Hence $g_1(x, \alpha) - g_2(x, \alpha) \equiv 0$. Q.E.D.

This uniqueness theorem must be used with caution. We have employed in its proof the fact that the only function $g(x, \alpha)$ satisfying the equation (v) and the condition $g(x_0, \alpha) = 0$ whenever $\alpha = \alpha_0 + m$, $m = 0, 1, 2, \ldots$, is identically zero. There are, on the other hand, functions which satisfy the equation (v) and the condition

$$g(x_0, \alpha) = 0, \qquad \alpha_0 \leqq \alpha \leqq \alpha_1, \qquad \alpha_1 \neq +\infty,$$

which do not vanish identically. For example, if

$$g(x, \alpha) \equiv \Gamma(\alpha) x^{-\alpha},$$

then

$$\frac{\partial}{\partial x} g(x, \alpha) = - g(x, \alpha+1) , \tag{6}$$

and

$$g(0, \alpha) = \begin{cases} 0, & \alpha < 0; \\ \infty, & \alpha \geqq 0. \end{cases}$$

Then, fortunately for Corollary [9.4], it cannot be used to prove that $0 = \Gamma(\alpha) z^{-\alpha}$, even though both sides of this false equation are solutions of the equation (6) and both satisfy the condition $g(0, \alpha) = 0$ when $\alpha < 0$. This example seems trivial, but in practice it is very easy to make false applications of Corollary [9.4] to less obvious cases unless one is careful. For example, consider the case when

$$F_1(z, \alpha) \equiv \sum_{n=0}^{\infty} \zeta(-\alpha-n) \frac{z^n}{n!}, \quad |z| < 2\pi, \quad \alpha \text{ not a positive integer;}$$

$$F_2(z, \alpha) \equiv \sum_{n=1}^{\infty} e^{nz} n^{\alpha}, \begin{cases} z < 0, & \alpha \text{ unrestricted;} \\ z = 0, & \alpha < -1, \end{cases}$$

where $\zeta(s)$ is Riemann's zeta function. $F_1(z, \alpha)$ and $F_2(z, \alpha)$ satisfy the F-equation, and the conditions

$$F_1(0, \alpha) = \zeta(-\alpha), \quad \alpha \text{ not a positive integer,}$$
$$F_2(0, \alpha) = \zeta(-\alpha), \quad \alpha < -1.$$

Are these two functions equal? Corollary [9.4] does not say yes, and it does not give us any way of saying no. As a matter of fact, it can be proved [see Lindelöf 1, pp. 138-139; Wirtinger 1; Truesdell 1, p. 149] that

$$F_2(z, \alpha) = e^{i(\alpha+1)\pi} \Gamma(\alpha+1) z^{-\alpha-1} + F_1(z, \alpha). \tag{7}$$

COROLLARY [9.5]. (REAL EXISTENCE THEOREM FOR THE F-EQUATION.) If $\phi(\alpha)$ is a given bounded function of α defined at least when $\alpha = \alpha_0 + m$,

$m = 0, 1, 2, \ldots$, then there exists a unique function $F(x, \alpha)$, defined at least when $\alpha = \alpha_0 + m$, $m = 0, 1, 2, \ldots$, which satisfies the F-equation and the boundary condition $F(x_0, \alpha) = \phi(\alpha)$. This function exists for every finite value of x. If $\phi(\alpha)$ is a continuous function of α when $\alpha \geqq \alpha_0$, then $F(x, \alpha)$ is also a continuous function of α for any real finite value of x when $\alpha \geqq \alpha_0$.

Proof: In Corollary [9.3], $C(x, \alpha) \equiv 1$. Hence $h = \min(a, 1)$, where for a we may choose any number whatever; then $h = 1$. Then by Corollary [9.3], $F(x, \alpha)$ exists (and is bounded) when $|x - x_0| < 1$. Now say $F(x_0 + 1/2, \alpha) = \phi_1(\alpha)$; this function exists when $|x - x_0 - 1/2| < 1$. By Corollary [9.3], in the range where $-1/2 < x - x_0 < 3/2$ (supposing for simplicity that $x_0 > 0$), these two solutions agree, so we may regard the one as an extension of the other. In this manner we may proceed until we reach any finite value of x. If $\phi(\alpha)$ is continuous when $\alpha \geqq \alpha_0$, all the conditions of Theorem [9.2] are satisfied, so $F(x, \alpha)$ is a continuous function of α. Q.E.D.

In our study of the F-equation we shall prefer complex variable existence and uniqueness theorems, which we shall give as Theorem [11.2] and Corollary [11.3].

§10. SOME OBSERVATIONS ON THE EXISTENCE THEOREMS

Our principal theorem [9.1] is no exception to the rule that existence theorems are but awkward means of deducing any results of formal interest. However, we shall consider one example in which the successive approximation procedure may be used.

Example [10.1]. If

$$f(x, \alpha) \equiv e^{-x} L_b^{(\alpha)}(x),$$

then $f(x, \alpha)$ satisfies the equation

$$\frac{\partial}{\partial x} f(x, \alpha) = - f(x, \alpha+1),$$

as may be verified from the formula (1.42). Now

$$f(0, \alpha) = \frac{\Gamma(\alpha+b+1)}{\Gamma(\alpha+1)\Gamma(b+1)} .$$

In the proof of Theorem [9.1] let $\vec{f}_0(x, \alpha)$ be $f(0, \alpha)$; then it is easy to compute the approximating functions $\vec{f}_r(x, \alpha)$ from the definition (A):

$$f_r(x, \alpha)$$

$$= \frac{\Gamma(\alpha+b+1)}{\Gamma(\alpha+1)\Gamma(b+1)} \sum_{m=0}^{r} (-)^m \frac{(\alpha+b+1)(\alpha+b+2)\dots(\alpha+b+m)}{(\alpha+1)(\alpha+2)\dots(\alpha+m)} \cdot \frac{x^m}{m!} .$$

Hence, letting r approach ∞, we see that

$$f(x, \alpha) = \frac{\Gamma(\alpha+b+1)}{\Gamma(\alpha+1)\Gamma(b+1)} \; {}_1F_1(\alpha+b+1; \alpha+1; -x) .$$

Recalling the definition (E.1) of Laguerre functions, we see that we have proved Kummer's first transformation (1.17) [Kummer 1, pp. 138-141; Copson 1, p. 261, ex. 2].

In §11 we shall indicate a much more efficient method of discovering and proving transformations of this type.

It is natural to begin to search for solutions of the F-equation, or of the equation (9.1), which satisfy very simple boundary conditions. We have seen that 0 is the only solution of the F-equation such that $F(z_0, \alpha) = 0$, $\alpha \geq \alpha_0$. The function $J_\alpha(x)$ satisfies the condition $J_\alpha(0) = 0$, $\alpha > 0$, but the function G in the appropriate equation (9.1) (i.e. equation (1.36)) does not satisfy the conditions of Theorem [9.1] in any region involving the point $z = 0$. We have already noticed that $e^{i\alpha\pi} z^{-1/2\alpha} J_\alpha(2\sqrt{z})$ satisfies the F-equation, but the boundary condition $F(0, \alpha) = e^{i\alpha\pi}/\Gamma(\alpha+1)$ satisfied by this solution is not very simple. As a matter of fact, any solution of an equation of the type (9.1) satisfying a boundary condition not essentially involving α is if the right side function is periodic a trivial one, as the following theorem shows.

THEOREM [10.1]. In the statement of Theorem [9.1], suppose further that

$$\text{(E)} \quad \vec{G}(\vec{t}, \vec{t}, \ldots, \vec{t}, x, \alpha+1) = \vec{G}(\vec{t}, \vec{t}, \ldots, \vec{t}, x, \alpha)$$

for all $\vec{t}$, x, α, and moreover that

$$\vec{\phi}(\alpha+1) = \vec{\phi}(\alpha), \quad \alpha = \alpha_0+m, \quad m = 0, 1, 2, \ldots.$$

Then when $(x, \alpha) \varepsilon W$,

$$\vec{f}(x, \alpha+1) = \vec{f}(x, \alpha),$$

and is moreover a solution of a certain system of ordinary differential equations.

Sketch of Proof: It is easy to show that under these conditions the functions $\vec{f}_r(x, \alpha)$ given by the definition (A) in the proof of Theorem [9.1] are periodic of period 1 in α if $\vec{f}_0(x, \alpha)$ is chosen to be periodic (say, $\phi(\alpha)$), and that hence so also is the solution $\vec{f}(x, \alpha)$. Then as far as this solution is concerned, the equation (9.1) takes the form

$$\frac{\partial}{\partial x}\vec{f}(x, \alpha) = \vec{G}(\vec{f}(x, \alpha), \vec{f}(x, \alpha), \ldots, \vec{f}(x, \alpha), x, \alpha),$$

$$= \vec{H}(\vec{f}(x, \alpha), x, \alpha),$$

say, and this is a system of ordinary differential equations.

THEOREM [10.2]. In the statement of Theorem [9.2], suppose further that $\vec{G}(\vec{t}, \vec{t}, \ldots, \vec{t}, x, \alpha)$ does not actually involve the argument α, and suppose $\vec{\phi}(\alpha) = \vec{A}$, where $\vec{A}$ is a constant vector. Then $\vec{f}(x, \alpha)$

does not involve α at all, and is the solution of a system of ordinary differential equations.

The proof is analogous to that of the preceding theorem.

We observe that neither essential assumption of these theorems can be weakened.

1. The periodicity of $\vec{G}(\vec{t}, \vec{t}, \ldots, \vec{t}, x, \alpha)$ in α is not sufficient to insure that all solutions of the equation (9.1) be periodic also. For example, $b^{\alpha} e^{bx}$ is a solution of the F-equation, for which $\vec{G}(\vec{t}^{(n)}, \vec{t}^{(n-1)}, \ldots, \vec{t}^{(o)}, x, \alpha) \equiv t_1^{(1)}$, but this solution is periodic only if $b = 0$ or $b = 1$.
2. If $\vec{G}(\vec{t}, \vec{t}, \ldots, \vec{t}, x, \alpha)$ is not periodic in α the mere fact that a particular solution is periodic for a fixed value of x does not insure that that particular solution be periodic for other values of x. For example, a solution of the equation

$$\frac{\partial}{\partial x} f(x, \alpha) = \frac{1}{\alpha+1} f(x, \alpha+1),$$

for which $\vec{G}(\vec{t}^{(n)}, \vec{t}^{(n-1)}, \ldots, \vec{t}^{(o)}, x, \alpha) \equiv t_1^{(1)}/(\alpha+1)$, is $\Gamma(\alpha+1)x^{-\alpha/2} J_{\alpha}(2\sqrt{x})$, but this solution is periodic if and only if $x = 0$.
3. For the existence of periodic solutions it is not necessary that $\vec{G}(\vec{t}^{(n)}, \vec{t}^{(n-1)}, \ldots, \vec{t}^{(o)}, x, \alpha)$ be periodic when its $n + 1$ vector arguments are not the same. For example, consider the case when

$$\vec{G}(\vec{t}^{(n)}, \vec{t}^{(n-1)}, \ldots, \vec{t}^{(o)}, x, \alpha) \equiv (t_1^{(1)} - t_1^{(o)})\alpha.$$

When $t_1^{(1)} \neq t_1^{(0)}$ this function is not periodic in α, but nevertheless the equation

$$\frac{\partial}{\partial x} f(x, \alpha) = [f(x, \alpha+1) - f(x, \alpha)]\alpha$$

has the periodic solution $f(x, \alpha) \equiv 0$.

THEOREM [10.3]. If there exists a single solution $\vec{f}(x, \alpha)$ of the equation (9.1) such that $\vec{f}(x, \alpha+1) = \vec{f}(x, \alpha)$ whenever $(x, \alpha) \in W$, then for that particular function $\vec{f}(x, \alpha)$

$$\vec{G}(\vec{f}(x, \alpha), \vec{f}(x, \alpha), \ldots, \vec{f}(x, \alpha), x, \alpha)$$

$$= \vec{G}(\vec{f}(x, \alpha), \vec{f}(x, \alpha), \ldots, \vec{f}(x, \alpha), x, \alpha) .$$

Proof: Suppose that $\vec{f}(x, \alpha+1) = \vec{f}(x, \alpha)$ and that $\vec{f}(x, \alpha)$ is a solution of the equation (9.1). Now

$$\frac{\partial}{\partial x} \vec{f}(x, \alpha+1) = \vec{f}(x, \alpha) .$$

By the equation (9.1),

$$\frac{\partial}{\partial x} \vec{f}(x, \alpha) = \vec{G}(\vec{f}(x, \alpha), \vec{f}(x, \alpha), \ldots, \vec{f}(x, \alpha), x, \alpha),$$

$$\frac{\partial}{\partial x} \vec{f}(x, \alpha+1) = \vec{G}(\vec{f}(x, \alpha), \vec{f}(x, \alpha), \ldots, \vec{f}(x, \alpha), x, \alpha+1).$$

Q.E.D.

We notice that this theorem does not state that the condition (E) is a necessary one for the existence of periodic solutions, for the condition (E) refers to an arbitrary vector $\vec{t}$ while the condition of this theorem refers only to the particular solution in question. Thus the equation

$$\frac{\partial}{\partial x} f(x, \alpha) = \alpha f(x, \alpha+1)$$

does not satisfy the condition (E), yet it possesses a periodic solution $f(x, \alpha) \equiv 0$.

Chapter IV

METHODS OF TREATING SPECIAL FUNCTIONS BASED ON THE UNIQUENESS THEOREM FOR THE CONDITION $F(z_0, \alpha) = \phi(\alpha)$

§11. POWER SERIES SOLUTIONS

The F-equation admits a very simple power series solution, which is easier to set up than the power series solutions of the typical linear ordinary differential equations of applied mathematics. By means of this solution we may rephrase and prove anew our real existence and uniqueness theorem (Theorem [9.5]) for the F-equation:

THEOREM [11.1]. Suppose the real function $\phi(\alpha)$ is absolutely bounded:

(1) $$|\phi(\alpha_0 + m)| < M, \quad m = 0, 1, 2, \ldots ,$$

where α_0 is a real constant. Then the function $F(x, \alpha)$ defined by the real power series

(2) $$F(x, \alpha) = \sum_{n=0}^{\infty} \phi(\alpha + n)\frac{(x - x_0)^n}{n!}$$

exists and is a real analytic function of x in any finite interval, at least for those values of α which exceed α_0 by a nonnegative integer, and is the unique analytic function $F(x, \alpha)$ satisfying the F-equation and the condition

(3) $$F(x_0, \alpha) = \phi(\alpha); \quad \alpha = \alpha_0 + m, \quad m = 0, 1, 2, \ldots ,$$

Proof: If we put x equal to x_0 in the series (2), we see at once that it satisfies the boundary condition (3). By comparison with the exponential series, the series (2) is absolutely convergent for all finite values of x. It represents therefore an analytic function of x. Differentiation under the sign of summation is permitted:

$$\frac{\partial}{\partial x} F(x, \alpha) = \sum_{n=1}^{\infty} \phi(\alpha + n) \frac{(x - x_0)^{n-1}}{(n-1)!},$$

$$= F(x, \alpha).$$

The function $F(x, \alpha)$ therefore satisfies the F-equation. It remains only to prove it unique. Suppose a second analytic function $F_1(x, \alpha)$ satisfies the F-equation and the boundary condition (3). Then $F(x, \alpha) - F_1(x, \alpha)$ is an analytic function satisfying the F-equation and vanishing when $x = x_0$, $\alpha = \alpha_0 + m$, $m = 0, 1, 2, \ldots$. Hence, by the F-equation, its first derivative vanishes for these values of x and α, and continuing in this fashion we see that all derivatives vanish, so the analytic function $F(x, \alpha) - F_1(x, \alpha)$ is identically zero in an interval about $x = x_0$, when $\alpha = \alpha_0 + m$, $m = 0, 1, 2, \ldots$. Q.E.D.

A very similar argument proves the complex variable existence theorem.

THEOREM [11.2]. Suppose the complex function $\phi(\alpha)$ is absolutely bounded

$$|\phi(\alpha_0 + m)| < M, \quad m = 0, 1, 2, \ldots \tag{4}$$

where α_0 is a complex constant. Then the function $F(z, \alpha)$ defined by the power series

$$F(z, \alpha) \equiv \sum_{n=0}^{\infty} \phi(\alpha + n) \frac{(z - z_0)^n}{n!} \tag{5}$$

exists and is an integral function of the complex variable z, at least for those values of α which exceed α_0 by a nonnegative integer, and is the unique analytic function $F(z, \alpha)$ satisfying the F-equation and the condition

$$(6) \quad F(z_0, \alpha) = \phi(\alpha), \quad \alpha = \alpha_0 + m, \quad m = 0, 1, 2, \ldots .$$

We also have a strong uniqueness theorem which makes no use of the condition of boundedness of $\phi(\alpha)$. This theorem is the analogue for complex variables of Corollary [9.4].

COROLLARY [11.3]. There exists at most one function $F(z, \alpha)$ of the complex variable z, analytic in a domain containing the point $z = z_0$ when α has a given fixed value exceeding the complex constant α_0 by a nonnegative integer, which satisfies the F-equation and the boundary condition (6).

The proof utilizes the uniqueness portion of Theorem [11.2] in a fashion analogous to the way the proof of Corollary [9.4] utilizes the uniqueness portion of the proof of Corollary [9.3].

If the boundary condition (6) is strengthened to apply to a half plane of α, it is easy to see that the existence and uniqueness theorems may be extended so as to apply also in a half plane of α.

THEOREM [11.4]. In the statement of Theorem [11.2], let the condition (4) be replaced by the stronger condition

$$(7) \qquad |\phi(\alpha) < M, \quad R\alpha \geq \alpha_0 .$$

Then a solution $F(z, \alpha)$ of the F-equation such that

$$F(z_0, \alpha) = \phi(\alpha), \quad R\alpha \geq \alpha_0, \tag{8}$$

exists, is unique, is an integral function of z for each fixed value of such that $R\alpha \geq \alpha_0$, and is represented by the series (5).

COROLLARY [11.5]. There exists at most one function $F(z, \alpha)$ which is an analytic function of the complex variable z in a domain containing the point $z = z_0$, α being held at a fixed value such that $R\alpha \geq \alpha_0$, and which satisfies both the F-equation and the boundary condition (8).

This theorem is the fundamental theorem of this essay, as far as the applications are concerned.

To prove Theorem [11.4] and Corollary [11.5] one simply applies Theorem [11.2] and Corollary [11.3] to the totality of points in the strip $\alpha_0 + 1 \geq R\alpha \geq \alpha_0$.

We shall now prove that if a solution $F(z, \alpha)$ of the F-equation is a bounded function of α for a fixed value z_0 of z, then it is a function of exponential order in z.

COROLLARY [11.6]. Suppose that $F(z, \alpha)$ is a solution of the F-equation, and that there exists a value z_0 of z such that

$$|F(z_0, \alpha_0 + m)| < M, \quad m = 0, 1, 2, \ldots.$$

Then there exists a constant K, independent of m, such that, whatever the value of z,

$$|F(z, \alpha_0 + m)| < Ke^{|z|}, \quad m = 0, 1, 2, \ldots.$$

Proof: By Theorem [11.2],

$$F(z,\alpha_0+m) = \sum_{n=0}^{\infty} F(z_0,\alpha_0+n+m)\frac{(z-z_0)^n}{n!}, \quad m = 0, 1, 2, \ldots .$$

Hence

$$|F(z,\alpha_0+m)| \leqq \sum_{n=0}^{\infty} |F(z_0,\alpha_0+n+m)|\frac{|z-z_0|^n}{n!},$$

$$< Me^{|z-z_0|},$$

$$< Me^{|z_0|}e^{|z|}. \qquad \text{Q.E.D.}$$

A corresponding result, which we shall not state explicitly, holds with reference to a right half-plane of α.

The condition that the boundary function $\phi(\alpha)$ be bounded, which we have used in the preceding existence theorems, may be weakened, as the following theorem shows.

THEOREM [11.7]. Suppose the real function $\phi(\alpha)$ obeys the condition

$$\text{(9)} \qquad \sup_{n\to\infty} \left|\frac{\phi(\alpha_0 + n + 1)}{n\,\phi(\alpha_0 + n)}\right| = \frac{1}{k},$$

where $k \neq 0$. Then there exists a unique solution $F(x, \alpha)$ of the F-equation, satisfying the boundary condition (3) and representable by the series (2) when

$$|x - x_0| \leqq k - \delta \quad \text{and} \quad \alpha = \alpha_0 + m, \; m = 0, 1, 2, \ldots ,$$

for which values of x and α it is an analytic function of x.

Sketch of Proof: The condition (9) assures that when $|x - x_0| \leqq k - \delta$ the series (2) is uniformly and absolutely convergent. The steps of the proof of Theorem [11.1] then remain valid if properly modified.

There is a corresponding complex variable theorem, which we shall not state explicitly.

The conditions of Theorem [11.7] are sketched in Fig. 2. Let us consider first the discrete case, when $\phi(\alpha)$ is known only at the separate points $\alpha = \alpha_0 + m$, $m = 0, 1, 2, \ldots$, indicated by heavy dots. Then the $F(x, \alpha_0 + m)$ curves are determined within the limits $|x - x_0| < k$. The $F(x, \alpha_0 - m)$ curves, which are not drawn in the figure, may be determined only up to an arbitrary polynomial of degree m. Second, let us consider the continuous case, when $\phi(\alpha)$ is known when $\alpha \geq \alpha_0$, as indicated by the heavy curve. Then the $F(x, \alpha)$ surface is uniquely determined over the strip $|x - x_0| < k$, for all α to the right of α_0. There are an infinite number of sheets (not shown) of the $F(x, \alpha)$ surface when $\alpha < \alpha_0$, which all join on the $F(x, \alpha_0)$ curve.

It is interesting to notice that the series (2) may be discovered by the ordinary calculus of operators. Since the F-equation has the form (3.10):

$$(10) \qquad (D_z - E_\alpha)\, F(z, \alpha) = 0,$$

it follows formally that

$$(11) \qquad F(z, \alpha) = e^{(z-z_0)E_\alpha}\, \phi(\alpha),$$

and this is the series (2). We have already remarked in §3 that if $G(z, \alpha) = e^{-z} F(z, \alpha)$, then

$$(12) \qquad (D_z - \Delta_\alpha)\, G(z, \alpha) = 0;$$

hence, formally,

$$(13) \qquad G(z, \alpha) = e^{(z-z_0)\Delta_\alpha}\, \phi(\alpha).$$

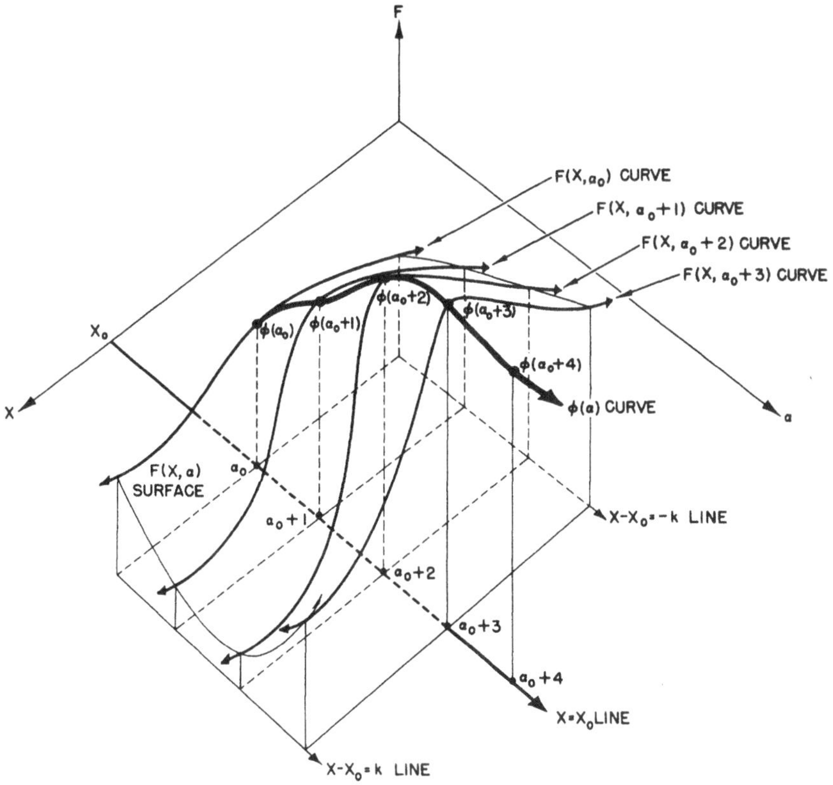

Sketch of real variable existence and uniqueness conditions for the equation $\frac{\partial}{\partial X} F(X,a) = F(X,a+1)$

1. Discrete conditions: If the boundary condition $F(X_0,a) = \phi(a)$ for $a = a_0 + i, i = 0, 1, 2, \ldots,$ is applied, the $F(X, a_0 + i), i = 0, 1, 2, \ldots,$ curves are uniquely determined, but the $F(X, a_0 - i)$ curves (not drawn) may be determined only up to an arbitrary polynomial of degree i in X.

2. Continuous conditions: If $\phi(a)$ is given for $a \geqq a_0$ (the $\phi(a)$ curve), then the $F(X,a)$ surface is uniquely determined for $a \geqq a_0$. There is an infinite number of sheets (not shown) of the $F(X,a)$ surface when $a < a_0$, which all join on the $F(X,a_0)$ curve.

FIGURE 2

We are thus led to the following theorem:

THEOREM [11.8]. The unique solution $F(x, \alpha)$ of the F-equation such that $F(x_0, \alpha) = \phi(\alpha)$, $\alpha = \alpha_0 + m$, $m = 0, 1, 2, \ldots,$ is given by the series

$$(14) \qquad F(x, \alpha) = e^{(x - x_0)} \sum_{n=0}^{\infty} \frac{(x - x_0)^n}{n!} \Delta^n \phi(\alpha),$$

which is certainly convergent when $|x - x_0| < k$, k being given by the formula (9), and may be convergent in a greater interval.

Sketch of Proof: The series (14) is easily shown to be a formal solution of the F-equation, and it clearly satisfies the boundary condition $F(x_0, \alpha) = \phi(\alpha)$. By Corollary [9.4] it is therefore the unique solution of the F-equation satisfying this condition. Convergence may be discussed analogously to the convergence of the Euler transformation of a series [Knopp 1, pp. 244-246, 262-265, 270-271, 468-470, 509-518].

The formula (14) has been given by Doetsch [Doetsch 1, p. 259].

COROLLARY [11.9]. If the series on the left is convergent, then

$$(15) \qquad e^{-x} \sum_{n=0}^{\infty} a_n \frac{x^n}{n!} = \sum_{n=0}^{\infty} \Delta^n a_0 \frac{x^n}{n!} .$$

Proof: By Corollary [9.4], we may equate the series (2) and (14). If we define a_n as $\phi(\alpha + n)$, the formula (15) follows. Q.E.D.

The formula (15) is a familiar generalization of the Euler transformation of a series. [Knopp 1, p. 270, ex. 116.

Knopp's difference operator Δ is the negative of ours, so his formula differs from ours by containing $(-)^n \Delta^n a_0$ instead of $\Delta^n a_0$.]

In a similar fashion we can easily see that the unique solution $F(x, \alpha)$ of the F-equation, satisfying the boundary condition $F(x_0, \alpha) = \phi(\alpha)$, is given formally by the series

$$(16) \qquad F(x, \alpha) = e^{-(x-x_0)} \sum_{n=0}^{\infty} \frac{2^n (x-x_0)^n}{n!} \mathbf{M}^n \phi(\alpha),$$

where $\mathbf{M}$ is the mean operator. Comparison of this result with the series (2) and (14) leads to other transformations of power series similar to the formula (15). Substitution of familiar functions in the series (14) and (16) leads to numerous formulas of the type discussed in the standard treatments of transformations of series [e.g. Knopp 1, pp. 270-271, ex. 115-123] so we shall not carry the matter further here. The series (5), because of its simplicity, can be used with some profit, as the following examples show. In particular, we obtain a complete answer to Question 4 of §1.

Example [11.1]. If in the formula (5) we substitute solution no. 26 of §5, using the boundary value (E.2), we find that

$$e^{-z} L_b^{(\alpha)}(z) = \sum_{n=0}^{\infty} \frac{\Gamma(\alpha+n+b+1)}{\Gamma(b+1)\Gamma(\alpha+n+1)} \cdot \frac{(-z)^n}{n!} .$$

With the aid of the definition (E.1) we may write this formula either in the form

$$(17) \qquad e^{-z} L_b^{(\alpha)}(z) = \frac{\sin b\pi}{\sin(\alpha+b)\pi} L_{-\alpha-b-1}^{(\alpha)}(-z)$$

or in the form (1.17). Thus Kummer's first transformation, which we have already proved once before as our formula (10.1), is a simple special case of our general expansion (5). If we substitute solution no. 30 of §5 into the formula (5), we obtain with equal ease Kummer's second transformation (1.18) [Kummer 1, pp. 138-141; Copson 1, p. 261, ex. 3]. If we substitute solution no. 37 of §5 into the formula (5), we discover the power series (1.19) for the Bessel function $J_a(y)$. Finally, if we substitute solution no. 5 of §5 into the formula (5) we obtain Euler's transformation (1.20).

Naturally one must choose between on the one hand using the Euler and Kummer transformations to find new solutions of the F-equation, or on the other using the fact that certain functions satisfy the F-equation to discover and prove the transformations. The key to our attitude in this essay lies in the word "discover." In the twentieth century we have no need to discover Euler's and Kummer's transformations, since they are now well known, and the method we choose to employ in proving them in text books is not very important. If we did not know the forms of these transformations, however, we should probably find it easier to discover them by observing that the simple functions they involve are solutions of differential·difference equations, methodically reducing these equations to the F-equation, and then automatically obtaining power series for them, than to have to grope towards conjecturing the transformations themselves. The advantage of the theorems of this essay is that they show one how to discover formulas without more than a very vague notion of what they will be. In the present case, for example, giving ourselves the problem "find a power series for $(1-z)^{-b} F(b;c;a;\frac{z}{z-1})$" we automatically discovered the Euler transformation. The essence of all these transformations from our present point of view may be stated thus: If $f(y, \alpha)$ is a function of interest, and if $L(z, \alpha) \cdot f(h(z), \alpha)$ is a solution of the F-equation, then there exists a "transformation"

$$(18) \qquad L(z, \alpha)\, f(h(z), \alpha) = \sum_{n=0}^{\infty} A_n (z-z_0)^n,$$

where $A_n = L(z_0, \alpha+n)\, f(h(z_0), \alpha+n)/n!$. In the case of any particular function we have no trouble in finding explicitly the functions $L(z, \alpha)$ and $h(z)$ and the coefficients A_n, by the reductions of §3 and the formula (5) respectively, once we know a differential-difference equation of type (3.1) satisfied by $f(y, \alpha)$. For the Euler transformation, for example, the particular function is $F(b + \alpha, c; \alpha; x)$,

$$(19) \qquad \begin{aligned} h(z) &= \frac{z}{z-1}, \\ L(z, \alpha) &= \frac{\Gamma(\alpha+b)\Gamma(\alpha-c)}{\Gamma(\alpha)} (1-z)^{-\alpha-b}, \end{aligned}$$

and if $z_0 = 0$

$$(20) \qquad A_n = \frac{\Gamma(\alpha+b+n)\Gamma(\alpha+n-c)}{\Gamma(\alpha+n)}.$$

Knowing <u>only the differential-difference equation</u> (C.13) <u>and the fact that</u> $F(a, b; c; 0) = 1$, with the methods already outlined one is led <u>automatically</u> to the equations (19) and (20), and hence to Euler's transformation.

In a sense, then, we may exhaust the possibilities for transformations of the Euler-Kummer type by enumerating all solutions of the F-equation and substituting them in the formula (5). We have thus found a general class of formulas of which the transformations of Euler and Kummer are hitherto isolated uncoordinated special cases.

In subsequent examples we shall leave it to the reader to depict the generality, simplicity, and heuristic value of the theorems whose use is exemplified, contenting ourselves with simply outlining the formal steps necessary to carry out their application to the particular problem in question.

Out of the many possibilities we select two further illustrations of the use of the formula (5), the first because it leads to a more complicated type of transformation and the second because it illustrates the use of Theorem [11.2] for defining a function.

Example [11.2]. Consider solution no. 15 of §5:

$$F(z, \alpha) = \Gamma(\alpha-b+1)(z^2+1)^{-\frac{\alpha+1}{2}} P_\alpha^b\left(-\frac{z}{\sqrt{z^2+1}}\right).$$

From the formula (D.17) we conclude that

$$F(0, \alpha) = \frac{\cos\frac{\alpha-b}{2}\pi}{2^b\sqrt{\pi}} \cdot \frac{\Gamma(\frac{\alpha}{2}-\frac{b}{2}+\frac{1}{2})}{\Gamma(\frac{\alpha}{2}+\frac{b}{2}+1)}\Gamma(\alpha+b+1) .$$

Then by formula (5),

$F(z, \alpha)$

$$= \frac{1}{2^b\sqrt{\pi}}\sum_{n=0}^{\infty} \cos\frac{\alpha+n-b}{2}\pi \; \Gamma(\alpha+b+n+1)\frac{\Gamma(\frac{\alpha}{2}-\frac{b}{2}+\frac{n}{2}+\frac{1}{2})}{\Gamma(\frac{\alpha}{2}+\frac{b}{2}+\frac{n}{2}+1)} \cdot \frac{z^n}{n!} .$$

A simple rearrangement yields the formula

(21) $P_\alpha^b(\cos\theta)$

$$= 2^b \csc^{\alpha+1}\theta\left\{\frac{\Gamma(\frac{\alpha}{2}+\frac{b}{2}+\frac{1}{2})\cos\frac{\alpha-b}{2}\pi}{\Gamma(\frac{\alpha}{2}-\frac{b}{2}+1)\,\Gamma(\frac{1}{2})} F\left(\frac{\alpha}{2}+\frac{b}{2}+\frac{1}{2}, \frac{\alpha}{2}-\frac{b}{2}+\frac{1}{2}; \frac{1}{2}; -\cot^2\theta\right)\right.$$

$$\left. + \cot\theta\,\frac{\Gamma(\frac{\alpha}{2}+\frac{b}{2}+1)\sin\frac{\alpha-b}{2}\pi}{\Gamma(\frac{\alpha}{2}-\frac{b}{2}+\frac{1}{2})\,\Gamma(\frac{3}{2})} F\left(\frac{\alpha}{2}+\frac{b}{2}+1, \frac{\alpha}{2}-\frac{b}{2}+1; \frac{3}{2}; -\cot^2\theta\right)\right\}.$$

The late Professor Bateman pointed out to me that this result by a use of Euler's transformation (1.20) may be shown to be equivalent to the more familiar formula

(22) $P_\alpha^b(x)$

$$= \frac{2^b \Gamma(\frac{\alpha}{2} + \frac{b}{2} + \frac{1}{2})}{\Gamma(\frac{1}{2})\Gamma(\frac{\alpha}{2} - \frac{b}{2} + 1)} \cos \frac{\alpha-b}{2}\pi \; F(\frac{b}{2} - \frac{\alpha}{2}, \frac{b}{2} + \frac{\alpha}{2} + \frac{1}{2}; \frac{1}{2}; x^2)$$

$$+ \frac{2^b x \Gamma(\frac{\alpha}{2} + \frac{b}{2} + 1)}{\Gamma(\frac{3}{2})\Gamma(\frac{\alpha}{2} - \frac{b}{2} + \frac{1}{2})} \sin \frac{\alpha-b}{2}\pi \; F(\frac{b}{2} - \frac{\alpha}{2} + \frac{1}{2}, \frac{b}{2} + \frac{\alpha}{2} + 1; \frac{3}{2}; x^2).$$

Example [11.3]. We may use the series (5) to generalize the Poisson-Charlier function. Consider the solution no. 36 of §5,

$$F(z, \alpha) = e^z \psi_m(-\alpha, z).$$

The definition (E.12)(E.13) of $\psi_m(p, z)$ given by Doetsch is valid only when m and p are positive integers. When α is a negative integer and b is a positive integer we see from the formula (E.14) that [Doetsch 1, p. 263],

$$(23) \qquad F(o, \alpha) = \cos(b + \alpha)\pi \binom{b}{-\alpha} .$$

Let us define the function $p_b(\alpha, z)$ by the relation

$$F(z, \alpha) = \frac{z^{-\alpha}}{\Gamma(1-\alpha)} p_b(-\alpha, z),$$

where $F(z, \alpha)$ is that solution of the F-equation which satisfies the condition (23), whatever the values of α and b. Substitution in the formula (5) yields at once the result

$$(24) \; p_b(\alpha, z) = \cos(b+\alpha)\pi \; z^{-\alpha} \frac{\Gamma(b+1)}{\Gamma(b-\alpha+1)} \; {}_1F_1(-\alpha; b-\alpha+1; z) ,$$

which reduces to the usual explicit expression (E.12) for the Charlier polynomials [Doetsch 1, p. 257; Szegö 1, p. 34] when α is a positive integer. From the expression (24) and the definition (E.1) we see that [Szegö 1, p. 34]

$$(25) \qquad p_b(\alpha, z) = \Gamma(\alpha+1) \cos(b+\alpha)\pi \, z^{-\alpha} L_\alpha^{(b-\alpha)}(z) .$$

Kummer's first transformation (1.17) may be deduced by substituting the solution no. 35 of §5 in the formula (5) and comparing the result with the formula (25).

When a formal power series solution diverges, it can often be summed by Nörlund's method, as shown by the following theorem:

THEOREM [11.10]. If there exists a function $h(\alpha)$ such that the indicated limit exists, the unique solution $F(z, \alpha)$ of the F-equation such that $F(z_o, \alpha) = \phi(\alpha)$ is given by the formula

$$(26) \qquad F(z, \alpha) = \lim_{k \to 0+} \sum_{n=0}^{\infty} e^{-kh(\alpha+n)} \phi(\alpha+n) \frac{(z-z_o)^n}{n!},$$

the limit being independent of the choice of functions $h(\alpha)$.

Proof: Let $F_k(z, \alpha)$ be defined as the series on the right:

$$F_k(z, \alpha) \equiv \sum_{n=0}^{\infty} e^{-kh(\alpha+n)} \phi(\alpha+n) \frac{(z-z_o)^n}{n!} .$$

By hypothesis, it is possible to choose a function $h(v)$ such that for all positive values of k the series is convergent, and hence, in a suitable domain, uniformly convergent in z. It is, then, a solution of the F-equation, such that

$$F_k(z_o, \alpha) = e^{-kh(\alpha)} \phi(\alpha) .$$

By hypothesis, $\lim_{k \to 0+} F_k(z, \alpha)$ exists; hence

$$F(z_0, \alpha) = \lim_{k \to 0+} e^{-kh(\alpha)} \phi(\alpha) ,$$

$$= \phi(\alpha) .$$

Uniqueness and hence independence of the particular choice of the function $h(v)$ follows from Corollary [11.3]. Q.E.D.

It is not difficult to prove, by analogy to Nörlund's proofs of the existence of sums [Nörlund 1, pp. 47-52, pp. 69-76; Milne-Thomson 1, pp. 209-213, pp. 222-237], that the limit exists for various types of functions ϕ if the function h be properly selected. Further consideration of this subject we defer to a subsequent memoir, which shall particularly concern the application of the theory of the F-equation to the generalized hypergeometric functions.

§12. FACTORIAL AND NEWTON SERIES SOLUTIONS

In the calculus of finite differences, factorial series are much more convenient than power series; in the theory of the F-equation both types of series are about equally easy to handle. Few results of formal interest can be found with the aid of factorial series, since so few familiar functions are representable by them, but the following theorem is at our disposal, should we wish to find the coefficients in a factorial series solution.

THEORÉM [12.1]. Suppose $\phi(\alpha)$ is expansible in a factorial series:

(1) $$\phi(\alpha) = \sum_{n=0}^{\infty} \frac{n!a_n}{\alpha(\alpha+1)\ldots(\alpha+n)}, \quad R\alpha \geqq \alpha_0 .$$

Then the unique solution $F(z, \alpha)$ of the F-equation such that $F(z_0, \alpha) = \phi(\alpha)$ is given by the factorial series

$$(2) \qquad F(z, \alpha) = e^{z-z_0} \sum_{n=0}^{\infty} \frac{n!a_n(z-z_0)}{\alpha(\alpha+1)\ldots(\alpha+n)}, \quad R\alpha \geq \alpha_0,$$

where

$$(3) \qquad a_n(y) \equiv \sum_{m=0}^{n} \frac{(-y)^m a_{n-m}}{m!} .$$

Proof: By Theorem [11.2], there exists a unique solution $F(z, \alpha)$ of the F-equation such that $F(z_0, \alpha) = \phi(\alpha)$, since $|\phi(\alpha)| < 1$ if $R\alpha$ is sufficiently large.

Let $\phi(\alpha)$ be given by the series (1). From a theorem of Nörlund [Nörlund 2, pp. 188-189] it follows that there exists a function $f(w)$, analytic within the circle $|w-1| = 1$, such that, when $R\alpha$ is sufficiently large, $\phi(\alpha)$ is representable by a Laplace integral:

$$\phi(\alpha) = \int_0^1 w^{\alpha-1} f(w)dw.$$

Let b be the order of $f(w)$ on the circle $|w-1| = 1$, and let b' be the order there of the function

$$\frac{f(w) - f(0+)}{w} .$$

Then if $\alpha_0 \geq 0$, $\alpha_0 = b - 1$, while if $\alpha_0 < 0$, $\alpha_0 = b' - 1$.

Consider now the function $H(z, \alpha)$ defined by the integral

$$H(z, \alpha) \equiv \int_0^1 w^{\alpha-1} e^{(z-z_0)w} f(w)dw.$$

Clearly this integral exists and represents a differentiable function of z. Differentiating, we find that

$$\frac{\partial H(z,\alpha)}{\partial z} = \int_0^1 w^{\alpha} e^{(z-z_0)w} f(w)dw,$$

$$= H(z, \alpha+1),$$

so the integral represents a solution of the F-equation. Since $H(z_0, \alpha) = \phi(\alpha)$, it follows from Theorem [11.2] that $H(z, \alpha) = F(z, \alpha)$. Then

$$(4) \qquad F(z, \alpha) = \int_0^1 w^{\alpha-1} e^{(z-z_0)w} f(w)dw .$$

The function $e^{(z-z_0)w} f(w)$ is analytic within the circle $|w-1| = 1$. Then it follows from the investigations of Nörlund [Nörlund 2, pp. 188-189] that there exists a factorial series representing the function $F(z, \alpha)$. If c is the abscissa of convergence of this series, $b(z)$ the order of $e^{(z-z_0)w} f(w)$ on the circle $|w-1| = 1$, and $b'(z)$ the order there of

$$e^{(z-z_0)w} \frac{f(w) - f(0+)}{w},$$

then $c = b(z) - 1$ if $c \geq 0$, while $c = b'(z) - 1$ if $c < 0$. But since the order of a function is unaffected when that function is multiplied by an analytic function [Nörlund 2, p. 47], $b(z) = b$ and $b'(z) = b'$. Hence $c = \alpha_0$.

We have proved the existence of a factorial series solution; it remains only to find its formal expression. For this purpose it is more convenient to consider, not the F-equation, but the equation (3.11). We shall actually show that the function defined by the series

$$(A) \qquad G(z, \alpha) = \sum_{n=0}^{\infty} \frac{n!a_n(z-z_0)}{\alpha(\alpha+1)\dots(\alpha+n)}$$

if formally a solution of the equation (3.11), viz.:

(B) $$\underset{z}{D}\, G(z, \alpha) = \underset{\alpha}{\Delta}\, G(z, \alpha);$$

it will follow that the series (2) is the solution of the F-equation whose existence we have proved by means of the Laplace integral (4). Differentiating and differencing the series (A), we see that

$$\underset{z}{D}\, G(z, \alpha) = \sum_{n=0}^{\infty} \frac{n!a_n'(z-z_0)}{\alpha(\alpha+1)\dots(\alpha+n)} ;$$

$$\underset{\alpha}{\Delta}\, G(z,\alpha) = \sum_{n=0}^{\infty} \frac{n!a_n(z-z_0)}{(\alpha+1)(\alpha+2)\dots(\alpha+n+1)} - \sum_{n=0}^{\infty} \frac{n!a_n(z-z_0)}{\alpha(\alpha+1)\dots(\alpha+n)} ,$$

$$= \sum_{n=0}^{\infty} \frac{n!a_n(z-z_0)[\alpha-(\alpha+n+1)]}{\alpha(\alpha+1)\dots(\alpha+n+1)} ,$$

$$= - \sum_{n=0}^{\infty} \frac{n!a_{n-1}(z-z_0)}{\alpha(\alpha+1)\dots(\alpha+n)} .$$

Hence the series (A) will satisfy formally the functional equation (B) if

$$a_n'(z-z_0) = -a_{n-1}(z-z_0).$$

The condition $G(z_0, \alpha) = \phi(\alpha)$ implies that $a_n(0) = a_n$. We easily find the expression (3) for $a_n(x)$. Q.E.D.

We should notice the simple form of the integral (4), which naturally suggests that other integrals of a similar type will also lead to solutions of the F-equation. These generalizations we shall consider in §13 and §16.

THEOREM [12.2]. Suppose $\phi(\alpha)$ is expansible in a Newton series:

(5) $$\phi(\alpha) = \sum_{n=0}^{\infty} (-)^n a_n \binom{\alpha-1}{n}, \quad R\alpha \geq \alpha_0 .$$

Then the unique solution $F(z, \alpha)$ of the F-equation such that $F(z_0, \alpha) = \phi(\alpha)$ is given by the Newton series

$$(6)\qquad F(z, \alpha) = e^{z-z_0} \sum_{n=0}^{\infty} (-)^n a_n(z-z_0)\binom{\alpha-1}{n}, \quad R\alpha \geq \alpha_0,$$

where

$$(7)\qquad a_n(w) = \sum_{m=0}^{\infty} a_{n+m} \frac{(-w)^m}{m!} .$$

Idea of Proof: The existence of the Newton series and its abscissa of convergence may be found with the aid of the generating function of Nörlund [Nörlund 2, pp. 149-150] in exactly the same fashion in which we proved Theorem [12.1]. Verification that the formal expressions (6) and (7) which we have given for this Newton series are correct may be left to the reader.

Example [12.1]. Since

$$(8)\; F(1-\alpha,b;c;a) = \frac{\Gamma(c)}{\Gamma(b)} \sum_{n=0}^{\infty} (-)^n \frac{\Gamma(b+n)}{\Gamma(c+n)} a^n \binom{\alpha-1}{n}, \quad |a| < 1,$$

it follows that the solution of the F-equation which reduces to $F(1-\alpha, b; c; a)$ when $z = 0$ is given by the series (6), where

$$(9)\qquad a_n(w) = \sum_{m=0}^{\infty} \frac{\Gamma(c)\Gamma(b+n+m)}{\Gamma(b)\Gamma(c+n+m)} a^{n+m} \frac{(-z)^m}{m!} ,$$

$$= \frac{\Gamma(c)\Gamma(b+n)}{\Gamma(b)\Gamma(c+n)} a^n \, {}_1F_1(b+n; c+n; -az) .$$

Then by Theorems [10.1] and [11.2],

$$(10)\quad \sum_{n=0}^{\infty} F(1-\alpha-n,\ b;c;a)\,\frac{(z-z_0)^n}{n!}$$

$$= \frac{e^{z-z_0}\Gamma(c)}{\Gamma(b)} \sum_{n=0}^{\infty} (-a)^n \frac{\Gamma(b+n)}{\Gamma(c+n)}\ {}_1F_1(b+n;\ c+n;\ -a(z-z_0))\ .$$

Using Kummer's first transformation (1.17) and the definition (E.1) of the Laguerre functions we may put this result in the form

$$(11)\quad \sum_{n=0}^{\infty} F(1-\alpha-n,\ b;c;a)\,\frac{(z-z_0)^n}{n!}$$

$$= \frac{\Gamma(c)\Gamma(b-c+1)}{\Gamma(b)} e^{(1-a)(z-z_0)} \sum_{n=0}^{\infty} (-a)^n \binom{\alpha-1}{n} L_{b-c}^{(c+n-1)}(a(z-z_0)).$$

The special case of this formula which we obtain by putting a equal to 1 yields, after some rearrangements a simple relation between the Laguerre functions:

$$(12)\qquad L_c^{(a+b)}(w) = \frac{\sin \pi c}{\sin \pi(c+a)} \sum_{n=0}^{\infty} (-)^n \binom{a}{n} L_{a+c}^{(c+b)}(w)\ .$$

When both a and c are positive integers we obtain the following formula for the Laguerre polynomials:

$$(13)\qquad L_p^{(m+b)}(w) = (-)^m \sum_{n=0}^{m} (-)^n \binom{m}{n} L_{m+p}^{(n+b)}(w)\ .$$

§13. CONTOUR INTEGRAL SOLUTIONS

In this section we shall answer Question 5 of §1 by showing that for a class of solutions of the F-equation there exist representations as contour integrals.

THEOREM [13.1]. Suppose $F(z,\ \alpha)$ is the solution of the F-equation such that

$$F(o, \alpha) = \phi(\alpha),$$

and suppose

$$\int_0^\infty e^{-zt} \; F(t, \alpha)dt$$

is an absolutely convergent integral if $Rz > k > 0$. Suppose $F(z, \alpha)$ is a continuous function of z when $z > 0$. Then, when $z > 0$,

$$(1) \quad F(z, \alpha) = - \frac{1}{2\pi i} \int_{c-i\infty}^{c+i\infty} e^{zw} w^{\alpha} \{ \mathop{S}_{\infty}^{\alpha} w^{-v-1} \phi(v) \Delta v \} dw, \quad c > k$$

provided the indicated sum exists.

Proof: Our hypotheses insure the existence of the Laplace transform $G(z, \alpha)$:

$$(2) \qquad G(z, \alpha) \equiv \int_0^\infty e^{-zt} \; F(t, \alpha)dt, \quad Rz > k.$$

Furthermore, the complex inversion integral

$$(3) \qquad F(z, \alpha) = \frac{1}{2\pi i} \int_{h-i\infty}^{h+i\infty} e^{zw} \; G(w, \alpha)dw, \quad h > k,$$

is valid [Doetsch 2, p. 105] when $z > 0$. Integrating by parts in the formula (2), we see that

$$G(z, \alpha) = - \frac{1}{z} e^{-zt} \; F(t, \alpha) \Big|_0^\infty + \frac{1}{z} \int_0^\infty e^{-zt} \; F(t, \alpha+1)dt,$$

$$= \frac{1}{z} \phi(\alpha) + \frac{1}{z} G(z, \alpha+1).$$

From this first order difference equation it follows that $G(z, \alpha)$ is of the form

$$(4) \qquad G(z, \alpha) = z^{\alpha}[\pi(\alpha, z) - \mathop{S}_{\alpha_0}^{\alpha} z^{-v-1} \; \phi(v)\Delta v],$$

where $\pi(\alpha+1, z) = \pi(\alpha, z)$. Now from the Laplace integral (2) it follows that

$$\lim_{z\to\infty} G(z, \alpha) = 0.$$

Hence $\pi(\alpha, z) = 0$, $\alpha_0 = \infty$, i.e.

$$G(z, \alpha) = -z^{\alpha} \mathop{S}_{\infty}^{\alpha} z^{-v-1}\phi(v)\Delta v .$$

By hypothesis, this sum exists. Substituting this formula for $G(z, \alpha)$ back in the complex inversion integral (3) yields the desired result. Q.E.D.

We should notice that in most cases of interest the simple series for the sum converges for sufficiently large values of z:

$$G(z, \alpha) = \frac{1}{z} \sum_{n=0}^{\infty} \frac{\phi(\alpha+n)}{z^n} .$$

Hence

$$(5) \qquad \int_0^{\infty} e^{-zt} F(t, \alpha) = \sum_{n=0}^{\infty} \frac{F(0,\alpha+n)}{z^{n+1}} ,$$

$$(6) \qquad F(z, \alpha) = \frac{1}{2\pi i} \int_{h-i\infty}^{h+i\infty} e^{zw} \{ \sum_{n=0}^{\infty} \frac{F(0,\alpha+n)}{w^{n+1}} \} dw, \quad h > k .$$

Example [13.1]. If we substitute solution no. 37 of §5 into the formula (6) we obtain the integral

$$(7) \quad J_a(y) = \frac{(\frac{1}{2} y)^a}{2\pi i \Gamma(a+1)} \int_{h-i\infty}^{h+i\infty} \frac{1}{w} e^{\frac{1}{4} y^2 w} {}_1F_1(1;a+1;-\frac{1}{w}) dw, \quad h > 0.$$

If we use Kummer's first transformation (1.17) we may transform this integral into a special case of the relation between Bessel and Laguerre functions which we shall give later as formula (16.6), which also contains

Schläfli's contour integral (1.21) as a different special case.

While the contour integral (1) is quite general, it is not often convenient for the handling of special cases. We now show a direct method of discovering many of the familiar contour integrals which represent various special functions, starting with a given representation of the boundary function $F(z_0, \alpha)$.

THEOREM [13.2]. Let $\phi(\alpha)$ be a given analytic function of α. Suppose that it is possible to find an analytic function of α, $\psi_w(\alpha)$, and a contour C such that

$$\phi(\alpha) = \int_C \psi_w(\alpha)dw, \qquad R\alpha \geq \alpha_0. \tag{8}$$

Suppose there exists a solution $\psi_w(z, \alpha)$ of the F-equation, analytic in z near $z = z_0$, such that $\psi_w(z_0, \alpha) = \psi_w(\alpha)$. Then there exists a solution $F(z, \alpha)$ of the F-equation, analytic in z near $z = z_0$, such that $F(z_0, \alpha) = \phi(\alpha)$. That solution can be represented by the integral

$$F(z, \alpha) = \int_C \psi_w(z, \alpha)dw, \qquad R\alpha \geq \alpha_0. \tag{9}$$

If the contour C is a part of the real axis which touches a singularity of the integrand or extends to ∞, conditions must be added to insure the analyticity of the integral in the formula (9) and the permissibility of differentiating under the sign of integration.

Proof: It is permissible to differentiate under the sign of integration in formula (9), so that it is easy to verify that the integral is a formal solution of the F-equation. It is also an analytic function. When $z = z_0$

it reduces to $\phi(\alpha)$, by formula (8). Hence by Theorem [11.5] it is the unique such solution. Q.E.D.

COROLLARY [13.3]. Suppose, for those values of w which lie on C,

$$|\psi_w(\alpha)| < M, \qquad R\alpha \geq \alpha_0 .$$

Then the formula (9) of Theorem [13.2] becomes

$$(10) \qquad F(z, \alpha) = \int_C \sum_{n=0}^{\infty} \psi_w(\alpha+n) \frac{(z-z_0)^n}{n!} \, dw, \qquad R\alpha \geq \alpha_0,$$

for all values of z.

The proof follows easily with the aid of Theorem [11.2].

Without illustration this theorem and corollary seem trivial, but the succeeding examples will show how they enable us to discover contour integrals for special functions.

Example [13.2]. Suppose we wish to find a contour integral for the Bessel function $J_a(x)$, knowing to start with Hankel's integral (A.1) for the Gamma function. We use the solution no. 37 of §5:

$$F(z, \alpha) \equiv e^{i\alpha\pi} z^{-\alpha/2} J_\alpha(2\sqrt{z});$$

$$F(0, \alpha) = \frac{e^{i\alpha\pi}}{\Gamma(\alpha+1)},$$

$$= \frac{e^{i\alpha\pi}}{2\pi i} \int_{-\infty}^{(0+)} e^w w^{-\alpha-1} \, dw .$$

Then by formula (10),

$$e^{i\alpha\pi} z^{-\alpha/2} J_\alpha(2\sqrt{z}) = \frac{e^{i\alpha\pi}}{2\pi i} \int_{-\infty}^{(0+)} e^w \sum_{n=0}^{\infty} \frac{w^{-\alpha-n-1}}{n!} (-z)^n \, dw,$$

$$= \frac{e^{i\alpha\pi}}{2\pi i} \int_{-\infty}^{(0+)} e^w w^{-\alpha-1} e^{-z/w} \, dw.$$

Hence

$$(11) \qquad J_\alpha(2\sqrt{z}) = \frac{z^{\alpha/2}}{2\pi i} \int_{-\infty}^{(0+)} w^{-\alpha-1} \exp(w - \frac{z}{w}) dw$$

this formula is Schläfli's integral (1.21) for the Bessel function $J_\alpha(x)$. [Schläfli 1, p. 203; Whittaker and Watson 1, pp. 244-245.]

Example [13.3]. Proceeding in an exactly analogous fashion with the solution no. 26 of §5,

$$F(z, \alpha) \equiv e^{i\alpha\pi} e^{-z} L_b^{(\alpha)} (z),$$

$$F(0, \alpha) = \frac{e^{i\alpha\pi}\Gamma(\alpha+b+1)}{\Gamma(b+1)\, 2\pi i} \int_{-\infty}^{(0+)} w^{-\alpha-1} e^w \, dw ,$$

by formula (10) we easily show that

$$(12) \; e^{-z} L_b^{(\alpha)}(z) = \frac{\Gamma(\alpha+b+1)}{\Gamma(b+1)\, 2\pi i} \int_{-\infty}^{(0+)} e^w w^{-\alpha-1} (1 + \frac{z}{w})^{-\alpha-b-1} dw.$$

The exponential function in this formula suggests we use Kummer's first transformation (10.1); we thus obtain the familiar contour integral (1.25) [Copson 1, p. 269, ex. 20] for the Laguerre functions:

$$(13) \qquad L_b^{(\alpha)}(z) = \frac{\Gamma(\alpha+b+1)}{\Gamma(b+1)\, 2\pi i} \int_{-\infty}^{(0+)} (1 - \frac{z}{w})^b e^w w^{-\alpha-1} \, dw.$$

The previous two examples show a direct connection between the two contour integrals derived; both are deducible by a mechanical process from the fact that

$F(0, \alpha)$ in each case is expressible in terms of the Gamma function, which in turn has a contour integral representation.

It is now natural to say, The combination of Gamma functions can be expressed in terms of the Beta function, for which we have quite a different contour integral representation. Can we deduce other contour integrals for the Bessel and Laguerre functions in this way? The following two examples answer this question.

Example [13.4]. As in Example [13.2] let us use the solution no. 37 of §5:

$$F(0, \alpha) = \frac{e^{i\alpha\pi}}{\Gamma(\alpha+1)},$$

$$= \frac{e^{i\alpha\pi}}{\Gamma(\alpha+1-b)\Gamma(b)} B(\alpha+1-b, b),$$

$$= -\frac{1}{4\Gamma(\alpha+1-b)\Gamma(b)\sin(\alpha-b)\pi\sin b\pi} \int_A^{(1+,0+,1-,0-)} w^{\alpha-b}(1-w)^{b-1}\,dw,$$

by Pochhammer's integral (B.1). Then by formula (10),

$$e^{i\alpha\pi}z^{-\frac{1}{2}\alpha} J_\alpha(2\sqrt{z})$$

$$= -\frac{1}{4\Gamma(b)\sin(\alpha-b)\pi\sin b\pi} \int_A^{(1+,0+,1-,0-)} (zw)^{-\frac{\alpha-b}{2}} w^{\alpha-b} (1-w)^{b-1}$$

$$\sum_{n=0}^{\infty} \frac{(-)^n (\sqrt{zw})^{\alpha-b+2n}}{n!\Gamma(\alpha-b+n+1)}\,dw,$$

$$= -\frac{1}{4\Gamma(b)\sin b\pi\sin(\alpha-b)\pi} \int_A^{(1+,0+,1-,0-)} w^{\alpha-b}(1-w)^{b-1}(zw)^{-\frac{\alpha-b}{2}}$$

$$J_{\alpha-b}(2\sqrt{zw})\,dw.$$

Hence

(14) $J_\alpha(2\sqrt{z})$

$$= \frac{-e^{-i\alpha\pi} z^{\frac{1}{2}b}}{4\Gamma(b)\sin b\pi \sin(\alpha-b)\pi} \int_A^{(1+,0+,1-,0-)} w^{\frac{\alpha-b}{2}} (1-w)^{b-1} J_{\alpha-b}(2\sqrt{zw})\,dw.$$

If we take the special case of this formula when

$$\alpha - b = -\frac{1}{2}, \ 2\sqrt{z} = y, \ w = v^2,$$

we easily deduce Hankel's integral (1.22) [Hankel 2, pp. 476-482; Whittaker and Watson 1, pp. 303-304, 306-307]:

$$(15) \quad J_a(y) = \frac{\Gamma(\frac{1}{2}-a)(\frac{1}{2}y)^a}{2\pi i \ \Gamma(\frac{1}{2})} \int_A^{(1+,-1-)} (v^2-1)^{a-\frac{1}{2}} \cos yv \, dv.$$

Example [13.5]. We may proceed in exactly the same fashion with the solution no. 26 of §5, adding only a seasonable use of Kummer's first transformation (10.1). We obtain the result

(16) $L_b^{(\alpha)}(z)$

$$= \frac{e^{-i\alpha\pi}\Gamma(\alpha+b+1)}{4i\Gamma(c)\Gamma(\alpha+b+1-c)\sin c\pi \sin(\alpha-c)\pi} \int_A^{(1+,0+,1-,0-)} w^{\alpha-c}(1-w)^{c-1}$$

$$L_b^{(\alpha-c)}(zw)\,dw \ .$$

We see from the preceding four examples that Bessel and Laguerre functions have each two integral representations, one of Gamma-function type and one of Beta-function type. Of these, the two former and one accidentally simple special case of one of the latter are well known. We see that we can proceed in the same fashion and find contour integrals of two types for <u>any</u> solution of the F-equation which for some fixed value of z is directly or inversely proportional to a single Γ-function whose argument involves α linearly.

The solution no. 15 of §5, which involves Legendre functions, does not appear to fall into the category we have just discussed. When $z = 0$, however, it is proportional to a Beta-function, so we may proceed with it in the fashion illustrated in Examples [13.4] and [13.5]. We obtain an integral which may be modified in two different ways so as to give the Schläfli and Laplace integrals (1.23) and (1.24) [Schläfli 2, p. 4; Laplace 1, p. 41; Whittaker and Watson 1, pp. 303-304, 306-307, 312-314], but the analysis is rather cumbrous, so we do not reproduce it here.

§14. TAYLOR'S THEOREM. GENERATING FUNCTIONS

In this section we shall treat Questions 1 and 2 of §1.

Taylor's theorem takes the form of a generating expansion for solutions of the F-equations:

THEOREM [14.1]. If the function $F(z, \alpha)$ satisfies the F-equation and if $F(z + y, \alpha)$ possesses a Taylor series, then this series may be put into the form

$$F(z + y, \alpha) = \sum_{n=0}^{\infty} \frac{y^n}{n!} F(z, \alpha + n). \tag{1}$$

Proof: If $F(z, \alpha)$ satisfies the F-equation,

$$\begin{aligned} \frac{\partial}{\partial z^n}[F(z+y, \alpha)]\Big|_{y=0} &= F(z+y, \alpha+n)\Big|_{y=0}, \\ &= F(z, \alpha+n). \end{aligned}$$

The formula (1) is therefore the expression of Taylor's theorem.

COROLLARY [14.2]. If for some value z_0 of z, $|F(z_0, \alpha)| < M$ when $R\alpha \geqq \alpha_0$, then for all values of z and y the expansion (1) is valid, when $R\alpha \geqq \alpha_0$.

Proof: Under this hypothesis of boundedness of $F(z_0, \alpha)$, it follows from Theorem [11.2] that $F(z, \alpha)$ is an entire function of z for each fixed value of α such that $R\alpha \geqq \alpha_0$.

Example [14.1]. If we substitute the solution no. 37 of §5 in the formula (1) we obtain Bessel and Lommel's formula (1.10) directly [Lommel 1, p. 11; Sonine 1, p. 22; Watson 1, p. 140; Copson 1, p. 343, ex. 13]:

$$(2) \qquad (z+y)^{-\frac{\alpha}{2}} J_\alpha(2\sqrt{z+y}) = \sum_{n=0}^{\infty} \frac{(-y)^n}{n!} z^{-\frac{\alpha+n}{2}} J_{\alpha+n}(2\sqrt{z}).$$

By Corollary [14.2] this formula is valid for all values z and y. If instead we use the solution no. 41 we find that while the conditions of Corollary [14.2] are not satisfied, we may apply Theorem [14.1] when $|z| - |y| > 0$, thus obtaining Lommel's second expansion formula [Lommel 1, p. 12; Watson 1, p. 140; Copson 1, p. 343, ex. 13]:

$$(3) \qquad (z+y)^{\frac{\alpha}{2}} J_\alpha(2\sqrt{z+y}) = \sum_{n=0}^{\infty} \frac{y^n}{n!} z^{\frac{\alpha-n}{2}} J_{\alpha-n}(2\sqrt{z}).$$

The solutions nos. 38-40, 42-44 show us that in both of the formulas (2) and (3) we may replace $J_\alpha(w)$ by $Y_\alpha(w)$, $H_\alpha^{(1)}(w)$, or $H_\alpha^{(2)}(w)$, the resulting expansions being valid when $|z| - |y| > 0$.

Example [14.2]. The conditions of Corollary [14.2] are satisfied by solution no. 26 of §5. Hence by formula (1) we find that [Magnus and Oberhettinger 1, p. 85]

$$(4) \qquad e^{-y} L_b^{(\alpha)}(z+y) = \sum_{n=0}^{\infty} \frac{(-y)^n}{n!} L_b^{(\alpha+n)}(z) .$$

This formula may be expressed in terms of confluent hypergeometric functions with the aid of the definition (E.1):

$$(5) \quad e^{-y} {}_1F_1(-b;\alpha+1;z+y)$$

$$= \frac{\Gamma(\alpha+1)}{\Gamma(\alpha+b+1)} \sum_{n=0}^{\infty} \frac{(-y)^n}{n!} \frac{\Gamma(\alpha+b+n+1)}{\Gamma(\alpha+n+1)} {}_1F_1(-b;\alpha+n+1;z) .$$

The formula (5) is a generalization of Kummer's first transformation (1.17), to which it reduces when $z = 0$. We shall later discover another similar generalization in our formula (16.14).

If we substitute the solution no. 27 of §5 into the formula (1) we find an expansion easily rearranged into the form

$$(6) \qquad (1+y)^{\alpha} L_b^{(\alpha)}(z+zy) = \sum_{n=0}^{\infty} \binom{\alpha+b}{n} y^n L_b^{(\alpha-n)}(z) .$$

A special case of this formula is

$$(7) \qquad 0 = \sum_{n=0}^{m} (-)^n \binom{m}{n} L_{m-a}^{(a-n)}(y) .$$

If we substitute the solution no. 28 of §5 into the formula (1) we find an expansion which is readily simplified into the form

$$(8) \quad (1-y)^{-a-b-1} \exp\left(-\frac{wy}{1-y}\right) L_a^{(b)}\left(\frac{w}{1-y}\right) = \sum_{n=0}^{\infty} \binom{a+n}{n} y^n L_{a+n}^{(b)}(w),$$

valid when $|y| < 1$. This formula has recently been given by Pinney [Pinney 1, p. 54]. The special case $a = 0$ of this formula is Sonine's classical generating function (1.3) for the Laguerre polynomials [Sonine 1, p. 42; Copson 1, p. 269, ex. 21; Szegö 1, p. 97].

If we substitute the solution no. 29 of §5 into the formula (1) we find an expansion which easily reduces to the form

$$(9) \qquad (1-y)^a L_a^{(b)}\left(\frac{w}{1-y}\right) = \sum_{n=0}^{\infty} \binom{a+b}{n} y^n L_{a-n}^{(b)}(w) .$$

Example [14.3]. The solutions nos. 33 and 34 of §5 when substituted into the formula (1) yield the following generating expansions for Hermite functions:

$$(10) \qquad \exp(2wy-y^2)\, H_\alpha(w-y) = \sum_{n=0}^{\infty} \frac{y^n}{n!} H_{\alpha+n}(w),$$

$$(11) \qquad H_a\left(\frac{1}{2} w + \frac{1}{2} y\right) = \sum_{n=0}^{\infty} \binom{a}{n} y^n H_{a-n}(w),$$

the special case $\alpha = 0$ of the formula (10) is the classical generating function (1.4) for the Hermite polynomials.

Example [14.4]. If we substitute the solutions nos. 15, 17, 19, and 21 of §5 into the formula (1), after some simplification we obtain the four generating expansions

$$(12) \quad (t^2-2tx+1)^{-\frac{a+1}{2}} P_a^b\left(\frac{x-t}{\sqrt{t^2-2tx+1}}\right) = \sum_{n=0}^{\infty} \binom{a-b+n}{n} t^n P_{a+n}^b(x),$$

$$(13) \quad (t^2-2tx+1)^{\frac{a}{2}} P_a^b\left(\frac{x-t}{\sqrt{t^2-2tx+1}}\right) = \sum_{n=0}^{\infty} \binom{a+b}{n} (-t)^n P_{a-n}^b(x),$$

$$(14)\quad \left(1-t^2-\frac{2xt}{\sqrt{1-x^2}}\right)^{-\frac{a}{2}} P_b^a(x+t\sqrt{1-x^2}) = \sum_{n=0}^{\infty} \frac{t^n}{n!} P_b^{a+n}(x) ,$$

$$(15)\quad \left(1-t^2-\frac{2xt}{\sqrt{1-x^2}}\right)^{\frac{a}{2}} P_b^a(x+t\sqrt{1-x^2})$$

$$= \sum_{n=0}^{\infty} (-t)^n \binom{a-b-1}{n} \frac{\Gamma(b+a-n)}{\Gamma(b+a)} P_b^{a-n}(x) .$$

While all these expansions apparently are given here for the first time in their full generality, many special cases are familiar. If in formula (12) we put b equal to a and use the formula (D.15) and (D.18) after some reductions we obtain Gegenbauer's generalization [Gegenbauer 1, p. 433; Whittaker and Watson 1, p. 329; Szegö 1, p. 82].

$$(16)\qquad (t^2-2tx+1)^{-a} = \sum_{n=0}^{\infty} t^n C_n^a(x) ,$$

of the Chevalier d'Alouville de Louville's classical generating function (1.2) for the Legendre polynomials [Louville 1, p. 132; Whittaker and Watson 1, p. 302].

If we put a equal to 1 and b equal to 0 in the formula (12), we find that

$$(17)\qquad \frac{x-t}{(t^2-2tx+1)^{3/2}} = \sum_{n=0}^{\infty} (n+1)t^n P_{n+1}(x) .$$

If we add $2t$ times this formula to the classical formula (1.2), we obtain the expression

$$(18)\qquad \frac{1-t^2}{(t^2-2tx+1)^{3/2}} = \sum_{n=0}^{\infty} (2n+1)t^n P_n(x) ,$$

which is known [Whittaker and Watson 1, p. 332, ex. 23].

When $a + b$ is a positive integer the series on the right in formula (13) terminates. In particular, for the Legendre polynomials we obtain the symbolic formula

$$(19) \qquad (t^2-2tx+1)[P]^p \frac{x-t}{\sqrt{t^2-2tx+1}} \doteq [P-t]^p x \; .$$

The special case when $t = \frac{1}{x}$ in the formula (19) may easily be reduced to the known formula [Whittaker and Watson 1, p. 331, ex. 9].

$$(20) \qquad \sin^p \theta \, P_p(\sin \theta) = \sum_{r=0}^{p} (-)^r \binom{p}{r} \cos^r \theta \, P_r \, (\cos \theta) \; .$$

Among other interesting consequences of formula (19) is the following expression,

$$(21) \qquad P_m(x) = \frac{\sqrt{\pi}}{2^m \Gamma(m+\frac{1}{2})} \sum_{j=0}^{2m} \binom{2m}{j} \frac{d^m}{dx^m} [x^j P_{2m-j}(x)] \; ,$$

which may be obtained from it by putting t equal to x, differentiating m times, and using Rodrigues' formula (1.12). Symbolically the equation (21) may be written

$$(22) \qquad [P]^m \, x = \frac{\sqrt{\pi}}{2^m \Gamma(m+\frac{1}{2})} D^m [P+x]^{2m} \; .$$

The solutions nos. 16, 18, 20 and 22 of §5 show us that in any of the formulas (12), (13), (14), or (15) $P_a^b(w)$ may be replaced by $Q_a^b(w)$. For example

$$(23) \, (t^2-2tx+1)^{-\frac{a+1}{2}} Q_a^b\left(\frac{x-t}{\sqrt{t^2-2tx+1}}\right) = \sum_{n=0}^{\infty} \binom{a-b+n}{n} t^n \, Q_{a+n}^b(x).$$

The special case of this formula for which $a = 0$ and $b = 0$ was discovered by Didon, and is familiar in the literature [Didon 1, p. 378; Laurent 1, p. 390; Heine 1, p. 134; Whittaker and Watson 1, p. 321, ex. 2].

In the preceding four examples we have exhausted the 22 possibilities for generating functions of the type (1) for the functions $J_a(w)$, $Y_a(w)$, $H_a^{(1)}(w)$, $H_a^{(2)}(w)$, $H_a(w)$, $L_b^{(a)}(w)$, $P_b^{(a)}(w)$, and $Q_b^{(a)}(w)$. We have shown that our original formulas (1.2), (1.3), (1.4), (1.5) and (1.10) are included as special cases of these 22 generating functions, while the remaining formulas (1.1), (1.6), (1.7), (1.8), and (1.9) must be of a different type. Thus the formula (1.1) is not an analogue of the formula (1.2), whose proper counterpart for the Bessel functions is the less familiar formula (1.5).

Example [14.5]. If we apply Theorem [14.1] to the solution no. 23 of §5 we obtain a result which can easily be rearranged into the formula

$$(24)\quad (1-t)^{-\alpha-1-\frac{1}{2}b}\, C_b^{\alpha+1}\left(\frac{x}{\sqrt{1-t}}\right) = \sum_{n=0}^{\infty} \binom{\alpha+n}{\alpha} t^n\, C_b^{\alpha+n+1}(x).$$

The special case of this formula when $\alpha = -\frac{1}{2}$ with the aid of the formula (D.15) may be expressed in the form

$$(25)\quad (1-2y\sqrt{1-x^2})^{-\frac{1}{2}(b+1)}\, P_b\left(\frac{x}{\sqrt{1-2y\sqrt{1-x^2}}}\right) = \sum_{n=0}^{\infty} P_{b+n}^n(x)\,\frac{y^n}{n!}.$$

This generating expansion is of an entirely different type as far as the Legendre functions are concerned, from those given in the preceding examples, since the summation is performed both upon the subscripts and upon the superscripts.

Example [14.6]. If we apply Theorem [14.1] to the solution no. 35 of §4 we have at once Doetsch's formula [Doetsch 1, p. 259]

$$(26)\qquad \psi_\alpha(b,\ z+y) = \sum_{n=0}^{\infty} \frac{y^n}{n!}\,\psi_{\alpha+n}(b,\ z).$$

Putting α equal to zero and supposing that b is a positive integer we obtain Doetsch's generating function for the Charlier polynomials [Doetsch 1, p. 260]:

$$(27) \qquad (1 + \frac{y}{z})^{b} e^{-y} = \sum_{n=0}^{\infty} \frac{y^n}{n!} p_n(b, z) .$$

If we apply Theorem [14.1] to the solution no. 36 of §4, we find that

$$(28) \qquad e^y \psi_b(a, z+y) = \sum_{n=0}^{\infty} \frac{y^n}{n!} \psi_b(a-n, z) .$$

By putting a equal to zero we obtain the formula

$$e^{-z} \cos b\pi = \sum_{n=0}^{\infty} \frac{y^n}{n!} \psi_b(-n, z);$$

hence

$$(29) \qquad \psi_b(-n, z) = 0.$$

This result has been given by Doetsch for the case when b is a positive integer [Doetsch 1, p. 263].

While Theorem [14.1] has enabled us to discover, coordinate and generalize a good number of the generating expansions we discussed in §1, there remain others which must be approached by different paths.

THEOREM [14.3]. Suppose $F(z, \alpha)$ is a solution of the F-equation such that, for fixed values of α and z,

$$(i) \qquad \sup_{n \to \infty} \frac{F(z, \alpha+n+1)}{F(z, \alpha+n)} = \frac{1}{k}, \qquad k \neq 0.$$

Then if $|w| < k$, for those values of w for which also there exists a real number h, less than one, such that

$$(ii) \qquad |F(z + tw, \alpha)| < e^{hs} \quad \text{if} \quad t > t_o,$$

it follows that

$$\sum_{n=0}^{\infty} F(z, \alpha+n)w^n = \int_0^{\infty} e^{-t} F(z+tw, \alpha)\, dt \tag{30}$$

provided the series converges uniformly in z in a domain including the fixed point z.

Proof: Let us write

$$K_\alpha(z, w) = \sum_{n=0}^{\infty} F(z, \alpha+n)w^n . \tag{A}$$

The condition (i) insures that the series is absolutely convergent when $|w| < k$, and we are assuming it is also uniformly convergent in z in a domain including the fixed point z. Hence

$$\frac{\partial}{\partial z} K_\alpha(z, w) = \sum_{n=0}^{\infty} F(z, \alpha+n+1)w^n ,$$

$$= \frac{1}{w}[K_\alpha(z, w) - F(z, \alpha)] .$$

Hence there exists a function z_0, $z_0 = z_0(\alpha, w)$, such that

$$K_\alpha(z, w) = -\frac{1}{w}\int_{z_0}^{z} e^{\frac{z-v}{w}} F(w, \alpha)dv .$$

If in this integral we make the substitution

$$v = z + tw$$

we find that

$$K_\alpha(z, w) = \int_0^{\frac{z-z_0}{w}} e^{-t} F(z+tw, \alpha)dt . \tag{B}$$

Now from the series (A) it is clear that z_0 must be chosen in such a way that

$$\lim_{w \to 0} K_\alpha(z, w) = F(z, \alpha),$$

no matter how the limit is approached. If the function $K_\alpha(z, w)$ is expressed by the integral (B), however, this limit will vary unless z_0 is so chosen that the upper limit of integration is independent of w, that is, $z_0 = -\infty$. Our hypotheses assure us that the resulting integral converges and defined an analytic function of z.

Q.E.D.

Example [14.7]. By substituting the solution no. 37 of §5 into the formula (30), after a little rearrangement we obtain the formula

$$(31)\quad \sum_{n=0}^{\infty} (-)^n y^n J_{a+n}(w) = \frac{w}{2y} \int_0^{\infty} e^{-\frac{wt}{y}} (t+1)^{-\frac{a}{2}} J_a(w\sqrt{t+1})dt,$$

valid when $Rw > 0$ and $Ry > 0$. In this formula $J_a(y)$ may be replaced by $Y_a(y)$, $H_a^{(1)}(y)$, or $H_a^{(2)}(y)$; naturally the range of validity will not be the same.

The integrals we obtain by substituting familiar solutions in the formula (30) are not often easy to evaluate. However, if we know the sum of the series we can use it to evaluate the integral. Thus, for example, if we take the formula (31) when $a = 0$ and add to it the series we get by using the fact that $J_n(x) = (-)^n J_{-n}(x)$, we find that

$$\sum_{n=-\infty}^{\infty} y^n J_n(w) + J_0(w) = \int_0^{\infty} e^{-t} \{J_0(\sqrt{w^2-2tyw}) + J_0(\sqrt{w^2+\frac{2ty}{y}})\}dt.$$

Hence by Schlömilch's formula (1.1) [Schlömilch 1, p. 137; Whittaker and Watson 1, p. 355]

$$(32)\int_0^{\infty} e^{-t} \{J_0(\sqrt{w^2-2tyw}) + J_0(\sqrt{w^2+\frac{2tw}{y}})\}dt = J_0(w)+e^{\frac{1}{2}w(y-\frac{1}{y})};$$

and in particular,

$$(33) \qquad 2\int_0^\infty e^{-t}\, J_0(\sqrt{w^2-2itw})dt = J_0(w) + e^{iw}.$$

Example [14.8]. By substituting the solution no. 33 of §5 into the formula (30) we show that

$$(34) \qquad \int_0^\infty e^{-t(1-2zy)-t^2y^2}\, H_\alpha(z - ty)dt = \sum_{n=0}^\infty y^n\, H_{\alpha+n}(z).$$

Putting α equal to zero in this formula and evaluating the integral on the left, we find that

$$(35) \qquad \frac{\sqrt{\pi}}{2y}\exp[(\frac{1}{2y} - z)^2]\,\mathrm{erfc}(\frac{1}{2y} - z) = \sum_{n=0}^\infty y^n\, H_n(z)\ .$$

The formula (35) may now be seen to be the natural analogue for the Hermite polynomials of the formula (1.1) for the Bessel coefficients.

The expression we have given in Theorem [14.3] naturally suggests we consider the sum

$$\sum_{n=0}^\infty F(z, \alpha-n)y^n\ .$$

Its evaluation is not so easy, but we shall be able to accomplish it in three special cases by the following theorem.

THEOREM [14.4]. Let $F(z, \alpha)$ be a solution of the F-equation such that for a fixed value α_0 of α and for values of z in a domain D,

$$(i) \qquad \inf_{z\varepsilon D}\ \sup_{n\to\infty} \frac{F(z, \alpha_0-n-1)}{F(z, \alpha_0-n)} = \frac{1}{k}, \qquad k \neq 0\ .$$

I. If $z_1 \varepsilon D$ and z_1 is a zero of $F(z, \alpha_0)$:

(ii) $$F(z_1, \alpha_0) = 0,$$

then

$$(36)\quad \sum_{n=0}^{\infty} F(z,\alpha_0-n)y^n = \int_{z_1}^{z} e^{(z-w)y}F(w,\alpha_0+1)dw, \quad |y| < k, \; z\varepsilon D,$$

provided $F(z, \alpha_0 + 1) \not\equiv 0$.

If

(iii) $$\int_0^{\infty} F(z+t, \alpha_0)dt$$

exists, $F(x, \alpha_0)$ is bounded as $x \longrightarrow \infty$, and $F(z, \alpha_0+1) \not\equiv 0$, then

$$(37)\quad \sum_{n=0}^{\infty} F(z, \alpha_0-n)y^n = -\int_0^{\infty} e^{-ty}F(z+t,\alpha_0+1)dt, \quad |y| < k, \; z\varepsilon D.$$

If

(iv) $$F(z, \alpha_0+1) \equiv 0,$$

then

$$(38)\quad \sum_{n=0}^{\infty} F(z+w,\alpha_0-n)y^n = e^{zy} \sum_{n=0}^{\infty} F(w,\alpha_0-n)y^n, \quad |y| < k, w\varepsilon D, \; (z+w) \varepsilon D.$$

Proof: Let us write

$$(A)\quad K(z, y) \equiv \sum_{n=0}^{\infty} F(z, \alpha_0-n)y^n .$$

Clearly

$$(B)\quad K(z, 0) = F(z, \alpha_0) .$$

The hypothesis (i) permits term-by-term differentiation in the definition (A). Then

$$\frac{\partial}{\partial z} K(z, y) = y\, K(z, y) + F(z, \alpha_0+1) .$$

Hence if $F(z, \alpha_0+1) \neq 0$ when $z \in D$ there exists a function $z_0(y)$ such that

$$\text{(C)} \qquad K(z, y) = e^{zy} \int_{z_0}^{z} e^{-wy} F(w, \alpha_0+1)dw.$$

Hence

(D)K(z,y)

$$= F(z, \alpha_0) - e^{y(z-z_0)} F(z_0, \alpha_0) - y \int_0^{z_0-z} e^{-wy} F(z+w, \alpha_0)dw.$$

Case I. The formula (D) becomes, when $y = 0$,

$$K(z, 0) = F(z, \alpha_0) - F(z_0, \alpha_0) .$$

Let us choose z_0 equal to z_1. Then the requirement (B) is satisfied. The formula (C) now becomes the formula (36). Q.E.D.

Case II. The hypotheses (iii) permit us to let z_0 approach $z + \infty$ in the formula (D):

$$K(z, y) = F(z, \alpha_0) - y \int_0^{\infty} e^{-ty} F(z+t, \alpha_0)dt .$$

The requirement (B) is now satisfied. The form assumed by the integral (C) then is

$$K(z, y) = \int_{z+\infty}^{z} e^{(z-w)y} F(w, \alpha_0+1)dw ,$$

which is readily transformed into the formula (37). Q.E.D.

Case III. By formula (C),

$$K(z, y) = e^{(z-z_0)y} z_0(y), \; z + z_0 \in D, \; |y| < k .$$

Since $K(z, y)$ is defined by a power series in y, it is evident that a power series exists for $z_0(y)$:

$$z_0(y) = \sum_{n=0}^{\infty} a_n y^n .$$

Hence

$$\sum_{n=0}^{\infty} F(z, \alpha_0 - n) y^n = e^{(z-z_0)y} \sum_{n=0}^{\infty} a_n y^n .$$

Putting z equal to z_0, we see that

$$\sum_{n=0}^{\infty} F(z_0, \alpha_0 - n) y^n = \sum_{n=0}^{\infty} a_n y^n .$$

Hence $a_n = F(z_0, \alpha_0 - n)$, so that

$$\text{(E)} \qquad \sum_{n=0}^{\infty} F(z, \alpha_0 - n) y^n = e^{(z-z_0)y} \sum_{n=0}^{\infty} F(z_0, \alpha_0 - n) y^n .$$

If in the formula (E) we replace z by $z + z_0$, we obtain the formula (38). Q.E.D.

The formula (E) is due to Appell, who deduced from it a number of interesting consequences, but did not apply it to familiar functions other than the Hermite polynomials [Appell 1, p. 120]. In Examples [14.10] to [14.14] we shall see that it includes as special cases familiar and unfamiliar generating expansions for several other special functions.

We should remark in passing that in Case III $F(z, \alpha_0 - n)$ is a polynomial of degree n.

Example [14.9]. Consider the solution no. 26 of §5:

$$F(z, \alpha) \equiv e^{i\alpha\pi} e^{-z} L_b^{(\alpha)}(z) .$$

It is clear that

$$\int_0^{\infty} e^{-t} L_b^{(\alpha)}(z+t) dt$$

exists if $R\alpha > Rb$, or if b is a positive integer and α is unrestricted; the integrand is certainly bounded as

$t \longrightarrow \infty$. Hence condition (iii) of Theorem [14.4] is applicable, so formula (37) is valid:

$$(39) \qquad \sum_{n=0}^{\infty} (-)^n L_b^{(\alpha-n)}(z)y^n = \int_0^{\infty} e^{-(1+y)t} L_b^{(\alpha+1)}(z+t)dt .$$

When $b = p$ and $\alpha + 1 = p$, this formula becomes

$$\sum_{n=0}^{\infty} (-)^n L_p^{(-p-1-n)}(z)y^n = \frac{(-)^p}{n!} \int_0^{\infty} e^{-(1+y)t}(z+t)^p dt .$$

The integral on the right is elementary:

$$(40) \qquad \sum_{n=0}^{\infty} (-)^n L_p^{(-p-1-n)}(z)y^n = (-)^p \sum_{m=0}^{p} \frac{z^m}{m!(1+y)^{p-m+1}} .$$

By letting p go to infinity in this formula we easily see that

$$(41) \qquad \lim_{p \longrightarrow \infty} (-)^p(1+y)^{p+1} \sum_{n=0}^{\infty} (-)^n L_p^{(-p-1-n)}(z)y^n = e^{z(1+y)}.$$

By applying Taylor's theorem to the formula (40) we see that

$$(-)^n n! L_p^{(-p-1-n)}(z) = (-)^p \frac{d^n}{dy^n} \Big[\sum_{m=0}^{p} \frac{z^m}{m!(1+y)^{p-m-1}} \Big]_{y=0} ,$$

$$L_p^{(-p-1-n)}(z) = (-)^p \sum_{m=0}^{p} \frac{z^m \Gamma(p-m+n+1)}{\Gamma(n+1)m!(p-m)!} .$$

Hence

$$(42) \qquad L_p^{(-n)}(z) = (-)^p \sum_{m=0}^{p} \frac{\Gamma(n-m)}{\Gamma(n-p)} \cdot \frac{z^m}{m!(p-m)!} ,$$

if n is an integer greater than p. It is easy to see that this formula is still correct if n is any integer; in fact

$$(43) \qquad L_p^{(-p+n)}(z) = \sum_{m=p-n}^{p} \frac{(-z)^m \, n!}{m!(p-m)!(m-p+n)!}$$

if $n \leq p$. These formulas we shall need to use in Example [14.10].

Example [14.10]. Consider solution no. 27 of §5,

$$F(z, \alpha) = \frac{z^{-\alpha} L_b^{(-\alpha)}(z)}{\Gamma(b-\alpha+1)} .$$

If $-R\alpha < 0$, condition (ii) of Theorem [14.4] is satisfied if $z_0 = 0$. Hence, if $Ra > 0$, from formula (36) we may show that

$$(44) \quad \sum_{n=0}^{\infty} \frac{z^{a+n} L_b^{(a+n)}(z)}{\Gamma(a+n+b+1)} y^n$$

$$= \frac{1}{\Gamma(a+b)} \int_0^z e^{(z-w)y} \, w^{a-1} \, L_b^{(\alpha-1)}(w) dw .$$

It follows from Taylor's theorem that

$$(45) \quad L_b^{(a+n)}(z) = \frac{\Gamma(a+n+b+1) z^{-a-n}}{n! \, \Gamma(b+a)} \int_0^z (z-w)^n w^{a-1} L_b^{(a-1)}(w) dw,$$

$$Ra > 0.$$

Formula (45) is a special case of Koshliakov's formula (1.32), of which we shall later prove a generalization as our formula (15.15).

If $z \neq 0$, $b = p$, and $R\alpha > p + 1$, α not being a positive integer, condition (iii) of Theorem [14.4] is satisfied by solution no. 27 of §5. In this case, then, by formula (37),

$$(46)\quad \sum_{n=0}^{\infty} \frac{z^{-\alpha+n} L_p^{(-\alpha+n)}(z)}{\Gamma(p-\alpha+n+1)} y^n$$

$$= -\frac{1}{\Gamma(p-\alpha)} \int_0^{\infty} e^{-ty} (z+t)^{-\alpha-1} L_p^{(-\alpha-1)} (z+t)dt .$$

If $\alpha = b$, condition (iv) of Theorem [14.4] is satisfied by solution no. 27 of §5. Hence, by formula (38),

$$(47)\quad \sum_{n=0}^{\infty} (z+y)^{-\alpha+n} L_\alpha^{(-\alpha+n)} (z+w) \frac{y^n}{n!}$$

$$= e^{zy} \sum_{n=0}^{\infty} w^{-\alpha-n} L_\alpha^{(-\alpha+n)} (w) \frac{y^n}{n!} .$$

We can simplify this formula when $\alpha = p$, and $y = 0$, since, by formulas (40) and (41) of the previous example,

$$\left.\frac{w^{-p+n} L_p^{(-p+n)}(w)}{n!}\right|_{w=0} = \frac{(-)^{p-n}}{(p-n)!n!} ,$$

$$= \frac{(-)^p}{p!} (-)^n \binom{p}{n} .$$

Hence we have as a special case of a special case of formula (38) Deruyts' expansion (1.9) [Deruyts 1, p. 9; Bateman 2, p. 452; Copson 1, p. 269, ex. 23]:

$$(48)\qquad \sum_{n=0}^{\infty} z^{-p+n} L_p^{(-p+n)}(z) \frac{y^n}{n!} = \frac{(y-1)^p}{p!} e^{zy} .$$

Example [14.11]. Consider solution no. 29 of §5,

$$F(z, \alpha) = \Gamma(\alpha-b)(-z)^{-\alpha} L_{-\alpha}^{(b)} \left(\frac{1}{z}\right) .$$

If $z \neq 0$, condition (iii) of Theorem [14.4] is satisfied when $R\alpha > 1$, but α is not a positive integer. Then by formula (37), after a little rearrangement we may show that if $R\alpha > 1$ and $Rw \geq 0$,

$$(49)\ \sum_{n=0}^{\infty} \frac{L_{-\alpha+n}^{(b)}(y)w^n}{\Gamma(n+b-\alpha+1)} = -\frac{1}{\Gamma(b-\alpha)} \int_0^{\infty} e^{-wt}(1+t)^{-\alpha-1}L_{-\alpha-1}^{(b)}(\tfrac{y}{1+t})dt.$$

If we put w equal to zero we find that

$$(50)\quad L_{-\alpha}^{(b)}(w) = (\alpha-b) \int_0^1 t^{\alpha-1} L_{-\alpha-1}^{(b)}(wt)dt, \quad R\alpha > 1.$$

If $\alpha = 0$, then condition (iv) of Theorem [14.4] is fulfilled by solution no. 29 of §5. Hence by formula (38),

$$(51)\ \sum_{n=0}^{\infty} (z+w)^n L_n^{(b)}(\tfrac{1}{z+w})\frac{y^n}{\Gamma(b+n+1)} = e^{zy} \sum_{n=0}^{\infty} w^n L_n^{(b)}(\tfrac{1}{w})\frac{y^n}{\Gamma(b+n+1)}.$$

Since

$$w^n L_n^{(b)}(\tfrac{1}{w})\big|_{w=0} = \frac{(-)^n}{n!},$$

it follows at once as a special case of formula (51) that

$$(52)\qquad \sum_{n=0}^{\infty} L_n^{(b)}(\tfrac{1}{z})\frac{(zy)^n}{\Gamma(n+b+1)} = e^{zy} y^{-\frac{b}{2}} J_b(2\sqrt{y});$$

an obvious rearrangement yields Sonine's generating expansion (1.8) [Sonine 1, p. 42; Copson 1, p. 270, ex. 29; Szegö 1, p. 98]. To Humbert must be conceded priority [Humbert 1, p. 63] in observing that the most natural method of deducing the formula (52) employs Appell's formula (38).

Example [14.12]. Consider the solution no. 46 of §5,

$$F(z, \alpha) = \frac{\phi_{-\alpha}^{(c)}(z)}{(-\alpha)!}.$$

Since the ϕ-polynomials are defined only when the lower index is a positive integer, we may say $\phi_{-1}^{(c)}(z) \equiv 0$; then condition (iv) of Theorem [14.4] is satisfied. Substitution in formula (38) yields Milne-Thomson's generalization of a classical identity for the Bernouilli polynomials [Milne-Thomson 1, p. 125]:

$$(53) \qquad \sum_{n=0}^{\infty} \phi_n^{(c)}(z+w) \frac{y^n}{n!} = e^{zy} \sum_{n=0}^{\infty} \phi_n^{(c)}(w) \frac{y^n}{n!} ,$$

from which a number of the properties of ϕ-polynomials are deducible.

Example [14.13]. If we consider the solution no. 36 of §5:

$$F(z, \alpha) \equiv e^z \psi_b(-\alpha, z) ,$$

from formula (29) we see that $F(z, 1) = 0$, whatever the value of b, so the formula (38) is applicable when $\alpha = 0$. Hence we easily show that

$$(54) \qquad \sum_{n=0}^{\infty} \frac{y^n}{n!} p_b(n, z+w) = e^{\frac{zy}{z+y}} \sum_{n=0}^{\infty} \frac{p_b(n, w)}{n!} \left(\frac{wy}{z+w}\right)^n .$$

If in this formula we put w equal to zero, by using our expression (11.24) we find that

$$(55) \qquad \sum_{n=0}^{\infty} \frac{y^n}{n!} p_b(n, z) = e^y \left(1 - \frac{y}{z}\right)^b \cos b\pi .$$

Comparing this result with the formula (27), we obtain Doetsch's formula [Doetsch 1, p. 257]:

$$(56) \qquad (-)^m p_n(m, y) = (-)^n p_m(n, y) .$$

Example [14.14]. From solution no. 15 of §5 and a suitable transformation of the type (5.1) we may show that $F(z, \alpha)$ as given by the definition

$$F(z, \alpha) = \frac{\cos \pi(\alpha-m)}{\Gamma(1-\alpha+m)} (z^2-1)^{-\frac{\alpha-m}{2}} P^m_{m-\alpha}\left(-\frac{z}{\sqrt{z^2-1}}\right) ,$$

is a solution of the F-equation. Clearly $F(z, 1) = 0$, so we may use the formula (38). We obtain the result

$$(57)\qquad \sum_{n=0}^{\infty} \frac{(-y)^n}{(m+n)!} [(z+w)^2-1]^{\frac{m+n}{2}} P_{m+n}^m\left(-\frac{z+w}{\sqrt{(z+w)^2-1}}\right)$$

$$= e^{zy} \sum_{n=0}^{\infty} \frac{(-y)^n}{n!} (w^2-1)^{\frac{m+n}{2}} P_{m+n}^m\left(-\frac{w}{\sqrt{w^2-1}}\right) .$$

This identity may be simplified in a number of ways. For example, if we put w equal to zero and use the formula (D.17), by a simple change of variable we obtain the result

$$(58)\ \sum_{n=0}^{\infty} \frac{t^n}{(m+n)!} P_{m+n}^m(x)$$

$$= \frac{2^m}{\pi m!} e^{xt} (1-x^2)^{\frac{m}{2}} [\Gamma(m+\tfrac{1}{2})]^2 {}_2F_3(m+\tfrac{1}{2}, m+\tfrac{1}{2}; \tfrac{1}{2}, \tfrac{m}{2}+\tfrac{1}{2}, \tfrac{m}{2}+1; -\tfrac{t^2}{4}(1-x^2)).$$

The special case of this formula which results when $m = 0$ is the formula (1.6), which has been given recently by Rainville, who gives credit for it to Catalan, F. H. Jackson, and others [Rainville 1, p. 269].

We now see that our original formulas (1.6), (1.8), and (1.9) share with many other generating expansions the property of being special cases of special cases of the formula (38).

The generating series

$$\sum_{n=0}^{\infty} F(z, \alpha-n) \frac{t^n}{n!}$$

we shall be able to evaluate in terms of the sum of the series

$$\sum_{n=0}^{\infty} F(z, \alpha-n)\ t^n$$

by the following theorem, whether or not $F(z, \alpha)$ satisfies the F-equation.

THEOREM [14.5]. If $K_{z,\alpha}(y)$ is defined

$$\text{(A)} \qquad K_{z,\alpha}(y) = \sum_{n=0}^{\infty} F(z, \alpha-n)y^n, \qquad |y| < k,$$

then

$$(59) \qquad \sum_{n=0}^{\infty} F(z, \alpha-n) \frac{y^n}{n!} = \frac{1}{2\pi i} \int_{-\infty}^{(0+)} e^w K_{z,\alpha}(\tfrac{y}{w})\frac{dw}{w} .$$

Proof: If the series (A) is convergent when $|y| < k$, the series on the left in the formula (59) represents an integral function of y. Let the contour of integration in the integral on the right lie completely outside the region where $|w| \leq |y|/k$. Then we may expand $K_{z,\alpha}(y/w)$ by the series (A) and integrate term by term; we obtain as end result the series on the left in formula (59). Q.E.D.

Example [14.15]. Consider Deruyts' expansion (1.9). Here

$$\left.\begin{aligned} F(z, \alpha) &\equiv \frac{z^{-\alpha}L_p^{(-\alpha)}(z)}{\Gamma(p-\alpha+1)} , \\ K_{z,\alpha}(y) &= \frac{(y-1)^{\alpha}}{\alpha!} e^{zy} , \end{aligned}\right\} \text{ where } \alpha = p;$$

hence by formula (59)

$$\sum_{n=0}^{\infty} z^{-p+n}L_p^{(-p+n)}(z) \frac{y^n}{n!n!} = \frac{1}{2\pi i p!} \int_{-\infty}^{(0+)} e^{w+\frac{zy}{w}} (\tfrac{y}{w}-1)^p \, dw,$$

$$= \frac{1}{p!} \sum_{r=0}^{p} (-y)^r\binom{p}{r} \cdot \frac{1}{2\pi i} \int_{-\infty}^{(0+)} e^{w+\frac{zy}{w}} w^{-r-1} dw;$$

then by Schläfli's integral (1.21),

$$(60) \quad \sum_{n=0}^{\infty} z^{-p+n}L_p^{(-p+n)}(z)\frac{y^n}{n!n!} = \frac{1}{p!} \sum_{r=0}^{p} (-)^r\binom{p}{r}(\tfrac{y}{z})^{\frac{r}{2}} I_r(2\sqrt{zy}).$$

We now give two theorems concerning the series

$$\sum_{n=0}^{\infty} F(z, \alpha+n) \frac{y^n}{n!\Gamma(\alpha+n+1)} .$$

THEOREM [14.6]. Suppose $F(z, \alpha)$, a solution of the F-equation, has a representation by a Laplace transform:

$$F(z, \alpha) = e^{i\alpha\pi} \int_0^{\infty} e^{-tz} t^{\alpha} \phi(t)dt . \tag{A}$$

Then

$$\sum_{n=0}^{\infty} F(z,\alpha+n)\frac{y^n}{n!\Gamma(\alpha+n+1)} = e^{i\alpha\pi} \int_0^{\infty} e^{-tz} \left(\frac{t}{y}\right)^{\frac{\alpha}{2}} J_{\alpha}(2\sqrt{ty})\phi(t)dt. \tag{61}$$

Sketch of Proof: Substitution of the formula (A) on the left in formula (61) and interchanging integration and summation yields the right side of equation (61). (The fact that $F(z, \alpha)$ is a solution of the F-equation is used only to secure the specific form (A) for the Laplace transform.)

THEOREM [14.7]. Let $F(z, \alpha)$ be a solution of the F-equation whose only singularities are at the points z_i. Then when the series is absolutely convergent,

$$\sum_{n=0}^{\infty} F(z,\alpha+n)\frac{y^n}{n!\Gamma(\alpha+n+1)} = \frac{1}{2\pi i} \int_C e^w w^{-\alpha-1} F(z+\frac{y}{w}, \alpha)dw , \tag{62}$$

where C is a contour starting at $-\infty$, encircling once counterclockwise a region containing all the points $y/(z_i-z)$, and returning to $-\infty$.

Proof: Each side of this equation, regarded as a function of y and α, is a solution of the F-equation.

As $y \longrightarrow 0$, all the singularities of the integrand also approach the origin. (If our contour C did not originally include all the points z_1, then as y approached 0 the excluded singularities would cross over the contour.) When $y = 0$, each side reduces to $F(z,\alpha)/\Gamma(\alpha+1)$. The theorem then follows as a consequence of Corollary [11.5]. Q.E.D.

Example [14.16]. Consider solution no. 15 of §5:

$$F(z, \alpha) = \Gamma(\alpha-b+1)(z^2-1)^{-\frac{\alpha+1}{2}} P_\alpha^b\left(-\frac{z}{\sqrt{z^2-1}}\right) .$$

When $b = 0$ we have the formula (v. Example [18.2])

$$F(z, n) = (-)^n \int_0^\infty e^{-tz} t^n \, I_0(t) dt .$$

Hence by formula (61),

$$(63) \quad \sum_{m=0}^\infty (z^2-1)^{-\frac{n+m+1}{2}} P_{n+m}\left(-\frac{z}{\sqrt{z^2-1}}\right) \frac{y^m}{m!}$$

$$= \int_0^\infty e^{-tz} \left(\frac{t}{y}\right)^{\frac{n}{2}} J_n(2\sqrt{ty}) \, I_0(t) dt .$$

When $n = 0$ we may compare this result with Catalan's formula (1.6), which we have proved already as a special case of our formula (58), and conclude that if $Rz > 1$

$$(64) \quad \int_0^\infty e^{-tz} J_0(2\sqrt{ty}) \, I_0(t) dt = \frac{1}{\sqrt{z^2-1}} e^{-\frac{zy}{z^2-1}} I_0\left(\frac{y}{z^2-1}\right) .$$

If we substitute our same solution into the formula (62) we obtain the result

$$(65)\qquad \sum_{n=0}^{\infty} \frac{\Gamma(\alpha-b+n+1)}{n!\,\Gamma(\alpha+n+1)}\, y^n\, (z^2+1)^{-\frac{\alpha+n+1}{2}}\, P^b_{\alpha+n}\left(-\frac{z}{\sqrt{z^2-1}}\right)$$

$$= \frac{\Gamma(\alpha-b+1)}{2\pi i} \int_C e^w [(zw+y)^2+w^2]^{-\frac{\alpha+1}{2}}\, P^b_{\alpha}\left(-\frac{zw+y}{\sqrt{(zw+y)^2+w^2}}\right) dw\ ,$$

where C encloses the points $t/(i-z)$, $t/(-i-z)$. When α and b are both the same positive integer m, by comparison with our formula (58) we may conclude that

$$(66)\qquad \frac{1}{2\pi i} \int_C \frac{e^w w^m}{[(zw+y)^2+w^2]^{m+1/2}}\, dw$$

$$= \frac{\Gamma(m+\frac{1}{2})}{\sqrt{\pi}\, m!}\, e^{-\frac{zy}{z^2+1}} (z^2+1)^{\frac{m}{2}}\ {}_2F_3\left(m+\tfrac{1}{2}, m+\tfrac{1}{2}; \tfrac{1}{2}, \tfrac{m}{2}+\tfrac{1}{2}, \tfrac{m}{2}+1; -\tfrac{y^2}{4}(1-x^2)\right).$$

Further manipulation of this contour integral is possible, but we omit it here.

§15. CERTAIN DEFINITE INTEGRALS

In this section we shall try to answer Question 5 of §1 by showing that it is possible to evaluate certain definite integrals in simple terms when the integrand involves a solution of the F-equation.

THEOREM [15.1]. Suppose that $F(z, \alpha)$ is a solution of the F-equation such that

$$(\text{i})\qquad F(z_0, \alpha) = 0, \qquad \alpha_0 \leqq R\alpha \leqq \alpha_0 + 1\ .$$

$$(\text{ii})\qquad \int_0^{z_0} z^{\alpha+b}\, F(z, \alpha)\, dz$$

exists and represents an analytic function of α when $\alpha_0 \leqq R\alpha \leqq \alpha_0 + 1$, $\alpha_0 + Rb > -1$.

(iii) The function $F(z, \alpha)$ is a continuous function of z when

$$0 \leqq z \leqq z_0, \quad \alpha_0 \leqq R\alpha \leqq \alpha_0 + 1.$$

(iv)
$$\frac{e^{-i\alpha\pi} \int_0^{z_0} z^{\alpha+b+1} F(z, \alpha)dz}{\Gamma(\alpha+b+1)}$$

is a bounded function of α everywhere in the strip

$$\alpha_0 \leqq R\alpha \leqq \alpha_0 + 1.$$

Then

$$(1) \int_0^{z_0} z^{\alpha+b} F(z, \alpha)dz = e^{i\alpha\pi}\Gamma(\alpha+b+1) f(b), \quad R(\alpha+b) > -1.$$

Note: $f(b)$ may be evaluated by giving any convenient special value α_1 to α, provided $R\alpha_1 > \alpha_0$, as we shall see in the first example succeeding the proof.

Proof: Define $K(\alpha, b)$:

$$K(\alpha, b) \equiv \int_0^{z_0} z^{\alpha+b} F(z, \alpha)dz .$$

By condition (iii) the following integration by parts is permitted:

$$\int_0^{z_0} z^{\alpha+b}F(z, \alpha)dz = \frac{z^{\alpha+b+1}}{\alpha+b+1}F(z, \alpha)\Big|_0^{z_0} - \frac{1}{\alpha+b+1}\int_0^{z_0} z^{\alpha+b+1}F(z, \alpha+1)dz.$$

By condition (i), the term between limits vanishes. Hence

$$K(\alpha, b) = -\frac{1}{\alpha+b+1} K(\alpha+1, b) .$$

By summing this difference equation we find that

$$K(\alpha, b) = e^{i\alpha\pi}\,\Gamma(\alpha+b+1)\,\pi(\alpha, b)\ ,$$

where $\pi(\alpha, b)$ is a function such that $\pi(\alpha+1, b) = \pi(\alpha, b)$. Hence

$$\frac{e^{-i\alpha\pi}\int_0^{z_0} z^{\alpha+b}\,F(z, \alpha)dz}{\Gamma(\alpha+b+1)} = \pi(\alpha, b)\ .$$

The function on the left is an analytic function of α in the strip $\alpha_0 \leqq R\alpha \leqq \alpha_0 + 1$, by hypothesis (ii), and by hypothesis (iv) it is bounded in the strip; since it is periodic of period 1 in α, it is analytic and bounded everywhere in the α plane, and hence by Liouville's theorem it is independent of α. Hence $\pi(\alpha, b) = f(b)$.

Q.E.D.

Example [15.1]. Consider the solution no. 37 of §5,

$$F(z, \alpha) = e^{i\alpha\pi} z^{-\frac{\alpha}{2}} J_\alpha(2\sqrt{z})\ .$$

If $z_0 = \infty$ all the conditions of Theorem [15.1] are satisfied, provided $R(b + \frac{1}{2}\alpha - \frac{1}{4}) < 0$ and $R(\alpha + b) > -1$. We shall wish to use the value $\frac{1}{2}$ for α_0, so we may be sure our theorem will be correct if $-\frac{3}{2} < Rb < 0$ (its range of validity may be extended later by analytic continuation). Then

$$\int_0^\infty z^{\frac{\alpha}{2}+b} J_\alpha(2\sqrt{z})dz = \Gamma(\alpha+b+1)\ f(b)\ .$$

To find $f(b)$, put α equal to $\frac{1}{2}$. Then

$$\frac{1}{\sqrt{\pi}}\int_0^\infty z^b \sin 2\sqrt{z}\ dz = \Gamma(b + \tfrac{3}{2})\ f(b)\ .$$

This integral converges when $-1 < Rb < -\frac{1}{2}$; we find that

$$f(b) = \frac{\sin\pi(b+1)\,\Gamma(2b+2)}{\sqrt{\pi}\;\Gamma(b+\frac{3}{2})},$$

$$= \frac{1}{\Gamma(-b)} .$$

Hence we have the formula of Lipschitz and Weber [Lipschitz 1, p. 192; Weber 1, pp. 227-229; Sonine 1, p. 39; Schafheitlin 1, pp. 158-161; Hardy 1, p. 14; Whittaker and Watson 1, p. 383, ex. 32; Watson 1, pp. 391-392]

$$\int_0^\infty z^{\frac{\alpha}{2}+b} J_\alpha(2\sqrt{z})dz = \frac{\Gamma(\alpha+b+1)}{\Gamma(-b)},$$

or, perhaps more conveniently,

$$(2)\quad \int_0^\infty x^{-a+b} J_b(cx)dx = \frac{1}{c}\left(\frac{2}{c}\right)^{-a+b} \frac{\Gamma(\frac{1}{2}b+\frac{1}{2})}{\Gamma(a-\frac{1}{2}b+\frac{1}{2})}, R(a-\tfrac{1}{2}) > Rb > -1.$$

This is our original formula (1.26), which we now see to be a special case of a general theorem for solutions of the F-equation.

Example [15.2]. Carrying out the same process with the solution no. 33 of §5, we easily show that

$$(3)\qquad \int_0^\infty e^{-z^2} z^{\alpha+b} H_\alpha(-z)dz = \frac{e^{i\alpha\pi}\Gamma(\alpha+b+1)\Gamma(\frac{1}{2})}{2^{b+1}\Gamma(\frac{b}{2}+1)} .$$

We should notice that even if the lower limit of integration is not zero and conditions (i) and (iv) are not satisfied, we may obtain a formal expression for the integral of Theorem [15.1]:

$$(4) \quad \int_a^{z_0} z^{\alpha+b} F(z, \alpha)dz$$

$$= e^{i\alpha\pi}\Gamma(\alpha+b+1)[\pi(\alpha,b)+\overset{\alpha}{S}e^{-iv\pi}\frac{z_0^{v+b+1}F(z_0,v)-a^{v+b+1}F(a,v)}{\Gamma(v+b+1)}\Delta v].$$

The conditions under which the formula (4) is valid, and under which $\pi(\alpha, b) = f(b)$, we shall not investigate.

Integrals of the form

$$\int_0^\infty e^{-bt} F(t, \alpha)dt$$

may be evaluated as indicated in the proof of Theorem [13.1], but a more general and practical approach is indicated by the following corollary to Corollary [11.5].

COROLLARY [15.2]. Let $F(z, \alpha)$ be an analytic solution of the F-equation such that $F(z_0, \alpha) = \phi(\alpha)$, $R\alpha \geq \alpha_0$. Then for those values of z such that there exists a constant k, $0 < k < 1$, such that

$$(1) \quad |F((z-z_0)t + z_0, \alpha)| < e^{k(ct)^{\frac{1}{a}}}, \qquad c > 0, a > 0,$$

when t is a sufficiently large positive number, the formula

$$(5) \quad \int_0^\infty e^{-(ct)^{\frac{1}{a}}} t^{\alpha+b-1} F((z-z_0)t + z_0, \alpha)dt$$

$$= \frac{a}{c^{\alpha+b}} \sum_{n=0}^\infty \frac{\Gamma(a\alpha+ab+an)}{n!} \phi(\alpha+n) \left(\frac{z-z_0}{c}\right)^n, \quad R(\alpha+b) \geq \delta > 0,$$

is correct when $|z-z_0| < h$, where h is defined by the formula

(ii) $$\sup_{n\to\infty} \left|\frac{\phi(\alpha+n+1)}{\phi(\alpha+n)} n^{a-1}\right| = \frac{|c|}{h|a^a|}, \quad R\alpha \geq \alpha_0 .$$

Proof: The conditions given assure that both the integral and the series define analytic functions of z. Clearly both are solutions of the F-equation, and when $z = z_0$ both sides reduce to

$$\frac{a}{c^{\alpha+b}} \Gamma(a\alpha+ab)\ \phi(\alpha), \qquad R\alpha \geq \alpha_0 .$$

The equation (4) then follows from Corollary [11.5].Q.E.D.

Example [15.3]. Consider the solution no. 37 of §5:

$$F(z, \alpha) = e^{i\alpha\pi} z^{-\frac{\alpha}{2}} J_\alpha(2\sqrt{z}) .$$

Then

$$F(0, \alpha) = \frac{e^{i\alpha\pi}}{\Gamma(\alpha+1)} .$$

$F(zt, \alpha)$ is bounded as $t \to \infty$, so condition (i) is satisfied. The condition (ii) takes the form

$$\lim_{n\to\infty} a^a n^{a-2} = \begin{cases} 0, & a < 2, \\ 4, & a = 2, \\ \infty, & a > 2. \end{cases}$$

It follows from Corollary [15.2], then, that when $R(\alpha+b) \geq \delta > 0$,

(6) $$\int_0^\infty e^{-(ct)^{\frac{1}{a}}} t^{\frac{\alpha}{2}+b-1} J_\alpha(2\sqrt{zt})dt = \frac{az^{\frac{\alpha}{2}}}{c^{\alpha+b}} \sum_{n=0}^\infty \frac{\Gamma(\alpha a+ab+an)}{n!\Gamma(\alpha+n+1)}\left(-\frac{z}{c}\right)^n,$$

the expansion being valid for all z if $0 < a < 2$, but in case $a = 2$ valid only if $|z| < |c|/4$. We may attach formal significance to the series on the right even if $a > 2$. Doing so for the moment, and supposing $a = m$, if we use the multiplication theorem (A.5) we obtain the formula

$$(7)\quad \int_0^\infty e^{-(ct)^{\frac{1}{a}}}\, t^{\frac{\alpha}{2}+b-1}\, J_\alpha(2\sqrt{zt})dt$$

$$= \frac{mz^{\frac{\alpha}{2}}\Gamma(m\alpha+mb)}{c^{\alpha+b}\,\Gamma(\alpha+1)}\; {}_mF_1(\alpha+b,\alpha+b+\tfrac{1}{m},\alpha+b+\tfrac{2}{m},\ldots,\alpha+b+\tfrac{m-1}{m};\alpha+1;-\tfrac{zm^m}{c}) .$$

This formula can be shown to be correct for all integer values of m, where the generalized hypergeometric function is given by the definition (C.1). Various special cases of the formula (7) are familiar in the literature [Lipschitz 1, p. 190; Weber 1, pp. 227-229; Gegenbauer 1, p. 439; Hankel 3, pp. 467-469; Gegenbauer 2, pp. 344, 346; Heine 1, pp. 242-243; Sonine 1, p. 45; Beltrami 1, p. 203; Beltrami 2, p. 481; Pincherle 1, p. 141; Callandreau 1, p. 122; Hobson 1, p. 74; Macdonald 1, pp. 432-433; Cailler 1, p. 315; Hardy 1, p. 13; Whittaker and Watson 1, p. 364, ex. 1, 2, p. 382, ex. 25; Watson 1, pp. 384-387; Koshliakov 1, p. 156; Bateman 2, p. 403, ex. 2; Copson 1, pp. 341-342, ex. 7, 8; Szegö 1, p. 99]; our formulas (1.27), (1.28), (1.33), and (1.35), of course, are among them, and we have now shown that these integrals are of the type to be expected in dealing with solutions of the F-equation.

Example [15.4]. If we choose the solution no. 26 of §5, using the formula (5) as in the previous example we show that

$$(8)\quad \int_0^\infty e^{-zt-(ct)^{\frac{1}{m}}}\, t^{\alpha+b-1}\, L_d^{(\alpha)}(zt)dt$$

$$= \frac{m\Gamma(m\alpha+mb)\Gamma(\alpha+d+1)}{c^{\alpha+b}\,\Gamma(\alpha+1)\Gamma(d+1)}\; {}_{m+1}F_1(\alpha+b,\alpha+b+\tfrac{1}{m},\ldots,\alpha+b+\tfrac{m-1}{m},\alpha+d+1;$$

$$\alpha+1;-\tfrac{zm^m}{c}).$$

The special case of this formula in which $m = 1$ may be reduced to the form

$$(9) \qquad F(a,b;c;y) = \frac{1}{\Gamma(a)} \int_0^\infty e^{-t} t^{a-1} \, {}_1F_1(b;c;yt)dt \ .$$

The special case of the formula (9) in which $y = 1$, $a = c$, may be shown with the aid of formula (C.6) to be equivalent to Sonine's formula (1.29) [Sonine 1, p. 42; Copson 1, p. 270, ex. 24].

All our original integrals (1.27), (1.28), (1.29), (1.34), and (1.35) we have now shown to be special cases of the same formula (5), which we could also apply to numerous other familiar functions.

We now develop a third general integral formula.

COROLLARY [15.3]. Let $F(z, \alpha)$ be an analytic solution of the F-equation such that $F(z_0, \alpha) = \phi(\alpha)$, $R\alpha \geq \alpha_0$. Then if

$$\sup_{n \to \infty} \left| \frac{1}{n} \cdot \frac{\phi(\alpha+n+1)}{\phi(\alpha+n)} \right| = \frac{1}{h}, \quad h \neq \infty, \ R\alpha \geq \alpha_0 \ ,$$

the formula

$$(10) \qquad \int_0^1 t^{\alpha+b-1} (1-t^{\frac{1}{a}})^{ac-1} \, F((z-z_0)t+z_0, \alpha)dt$$

$$= a\Gamma(ac) \sum_{n=0}^{\infty} \frac{\Gamma(a\alpha+ab+an) \, \phi(\alpha+n)}{\Gamma(a\alpha+ab+ac+an)} \frac{(z-z_0)^n}{n!}$$

is valid when $|z-z_0| < h$, $R(\alpha+b) \geq \delta > 0$, $Rc \geq \varepsilon > 0$, $Ra > 0$, $R\alpha \geq \alpha_0$.

Sketch of Proof: The conditions of the corollary insure that both the series and the integral define analytic functions of z for the indicated ranges of the variables and constants. Clearly each side is a solution of the F-equation. When $z = z_0$, each side reduces to

$$\frac{a\Gamma(ca)\Gamma(a\alpha+ab)\phi(\alpha)}{\Gamma(a\alpha+ab+ac)}, \qquad R\alpha \geqq \alpha_0 .$$

The formula (10) is then a consequence of Corollary [11.5].

Example [15.5]. Again using the solution no. 37 of §5 as in Example [15.3], by substitution in the formula (10) we can show that, for all values of z,

$$\int_0^1 t^{\frac{\alpha}{2}+b-1}(1-t^{\frac{1}{a}})^{ac-1} J_\alpha(2\sqrt{zt})dt \tag{11}$$

$$= az^{\frac{\alpha}{2}}\Gamma(ca)\sum_{n=0}^{\infty}\frac{\Gamma(a\alpha+ab+an)}{\Gamma(a\alpha+ab+ac+an)\,\Gamma(\alpha+n+1)}\cdot\frac{(-z)^n}{n!} .$$

When $a = m$, with the aid of the multiplication formula (A.5) we may put this relation in the form

$$\int_0^1 t^{\frac{\alpha}{2}+b-1}(1-t^{\frac{1}{m}})^{mc-1} J_\alpha(2\sqrt{zt})dt \tag{12}$$

$$= \frac{az^{\frac{\alpha}{2}}\Gamma(m\alpha+mb)}{\Gamma(m\alpha+mb+mc)\,\Gamma(\alpha+1)}\, {}_mF_{m+1}(\alpha+b,\alpha+b+\tfrac{1}{m},\alpha+b+\tfrac{2}{m},\ldots,\alpha+b+\tfrac{m-1}{m};$$

$$\alpha+b+c,\alpha+b+c+\tfrac{1}{m},\alpha+b+c+\tfrac{2}{m},\ldots,\alpha+b+c+\tfrac{m-1}{m},\alpha+1;-z) .$$

If in this result we put m and b both equal to 1, we find that

$$\int_0^1 t^{\frac{\alpha}{2}}(1-t)^{c-1} J_\alpha(2\sqrt{zt})dt = \Gamma(c)z^{-\frac{c}{2}} J_{\alpha+c}(2\sqrt{z}) . \tag{13}$$

This formula is equivalent to Sonine's integral (1.30) [Sonine 1, p. 36; Whittaker and Watson 1, p. 382, ex. 23; Copson 1, p. 343, ex. 12].

Example [15.6]. Again using the solution no. 26 of §5 as in Example [15.4], by substitution in the formula (10) in the special case when $a = 1$ we can show that

$$(14) \qquad \int_0^1 t^{\alpha+b-1}(1-t)^{c-1}\, e^{-zt}\, L_a^{(\alpha)}(zt)dt$$

$$= \frac{\Gamma(c)\Gamma(\alpha+b)\Gamma(\alpha+a+1)}{\Gamma(\alpha+1)\Gamma(a+1)\Gamma(\alpha+b+c)}\, {}_2F_2(\alpha+b,\alpha+a+1;\alpha+b+c,\alpha+1;-z),$$

when $R(\alpha+b) > 0$, $Rc > 0$. If in this formula we apply Kummer's first transformation (1.17), after a little re-arrangement we show that

$$(15) \qquad \int_0^1 t^{b-1}(1-t)^{c-1}\, L_a^{(\alpha)}(zt)dt$$

$$= \frac{\Gamma(c)\Gamma(b)\Gamma(\alpha+a+1)}{\Gamma(\alpha+1)\Gamma(a+1)\Gamma(b+c)}\, {}_2F_2(-a,\ b;\ \alpha+1,\ b+c;\ z)\ ,$$

when $Rb > 0$, $Rc > 0$. The special case of formula (15) which results from putting b equal to $\alpha + 1$ yields Koshliakov's formula (1.32) [Koshliakov 1, p. 155; Bateman 2, p. 454, ex. 3; Copson 1, p. 270, ex. 25].

When $a = -b-c$, $Rb > 0$, $Rc > 0$, then formula(15) reduces to the relation

$$(16) \int_0^1 t^{b-1}(1-t)^{c-1} L_{-b-c}^{(\alpha)}(zt) = \frac{\sin \pi(b+c)}{\sin \pi b} \cdot \frac{\Gamma(1+\alpha-b-c)}{\Gamma(1+\alpha-b)} L_{-b}^{(\alpha)}(z),$$

this result being a generalization of our earlier formula (14.50), which may be obtained from it by putting c equal to one.

We have seen in the preceding four examples that the formulas (1.30) and (1.32) are very special cases of a general integral formula for solutions of the F-equation.

We may say that since the three formulas (1), (5), and (10) include as special cases all the formulas (1.26) through (1.29), Question 6 of §1 is largely answered.

The method of the preceding two corollaries may be generalized, however, into the following formal procedure. Suppose $F(z, \alpha)$ is a solution of the F-equation such that $F(z_0, \alpha) = \phi(\alpha)$. Suppose $h(w)$ and $g(\alpha)$ are related by the integral equation

$$(17) \qquad \int_C w^\alpha h(w)dw = g(\alpha) ,$$

where C is some contour. Then, formally,

$$(18) \quad \int_C w^\alpha h(w)F((z-z_0)w+z_0, \alpha)dw = \sum_{n=0}^{\infty} \phi(\alpha+n)g(\alpha+n)\frac{(z-z_0)^n}{n!} .$$

Example [15.7]. In formulas (17) and (18) let the function $h(t)$ be given by the definition

$$h(t) = t^{b-1} J_c(2\sqrt{yt}),$$

and let $F(z, \alpha)$ be solution no. 37 of §5:

$$F(z, \alpha) = e^{i\alpha\pi} z^{-\frac{\alpha}{2}} J_\alpha(2\sqrt{z}).$$

Substituting into the formula (17) and using Weber's formula (1.26), we find that

$$g(\alpha) = \frac{\Gamma(\alpha + b + \frac{1}{2})}{y^{\alpha+b}\Gamma(\frac{1}{2}c-\alpha-b)} .$$

Substituting this result into formula (17) we have at once the Struve-Weber-Sonine formula [Struve 1, p. 92; Weber 2, pp. 75-80; Sonine 1, pp. 51-52; Schafheitlin 1, pp. 161-178; Hardy 1, p. 14; Whittaker and Watson 1, p. 383, ex. 38; Watson 1, pp. 396-410]

$$(19) \qquad \int_0^\infty t^{\frac{\alpha}{2}+b} J_c(2\sqrt{yt})\, J_\alpha(2\sqrt{zt})dt$$

$$= \frac{z^{\frac{\alpha}{2}}\,\Gamma(\alpha+b+\frac{1}{2}c)}{y^{\alpha+b}\Gamma(\alpha+1)\Gamma(\frac{1}{2}c-\alpha-b)}\, F(\alpha+b+\tfrac{1}{2}c, \alpha+b-\tfrac{1}{2}c; \alpha+1; \tfrac{z}{y}).$$

We shall not investigate the region of validity of this formula, nor the evaluation of the integral outside this range, since our purpose is not to make a list of relationships among special functions but rather to sketch briefly certain easy, direct, and fruitful methods of discovery. For a full discussion of the integral formula (19), we refer the reader to the passage in Watson's Bessel Functions which we have already cited.

Starting with the integral (19), we could carry out a similar procedure and evaluate the integral

$$\int_0^\infty t^{\frac{\alpha}{2}+b} J_d(2\sqrt{wt})\, J_c(2\sqrt{yt})\, J_\alpha(2\sqrt{zt})dt ,$$

thereby generalizing some results of Sonine and Macdonald [Sonine 1, p. 46; Macdonald 2, pp. 142-143; Whittaker and Watson 1, pp. 383-384, ex. 38-39; Watson 1, pp. 411-413]. The details of this evaluation, along with an infinite number of further definite integrals of the same sort, I leave to the reader, who by now doubtless is convinced that with the aid of the theory of the F-equation the discovery and proof of a variety of integrals involving well known special functions is a purely mechanical process.

§16. RELATIONS AMONG VARIOUS SOLUTIONS OF THE F-EQUATION

We shall now attempt to answer Question 8 of §1.

Many of the results we have discussed in the previous sections utilize the idea of linear combination of solutions. In the theory of ordinary linear differential equations or difference equations any solution (except for certain "singular" solutions) of a homogeneous equation is one of a manifold of linear combinations of certain fundamental solutions. While a linear combination of solutions of the F-equation with coefficients periodic in α is also a solution of the F-equation, we have no uniqueness theorem which will place special emphasis on this fact, or enable us to characterize any sort of fundamental solution, the existence of which the great variety of particular functions satisfying the F-equation (v. §5) makes unlikely. We may, however, by generalizing the idea of linear combination, set up a method of discovering and verifying numerous relationships between various particular solutions.

THEOREM [16.1]. Suppose $F(z, \alpha)$ is a solution of the F-equation, and suppose the functions $F_y(z, \alpha)$ form a set of solutions of the F-equation. Let O_y be an operator which

(i) binds the variable y; and

(ii) commutes, in the formula (2) below, with D_z, E_α, and the operation of replacing z by z_0.

Suppose, for some value z_0 of z,

$$F(z_0, \alpha) = O_y[F_y(z_0, \alpha)], \qquad R\alpha \geq \alpha_0. \tag{1}$$

Then, for all values of z such that the expression on the right has a meaning

$$F(z, \alpha) = O_y[F_y(z, \alpha)], \qquad R\alpha \geq \alpha_0. \tag{2}$$

Note: In future uses of the symbol O_y it will be supposed that O_y is defined only in cases where it has the commutability property used in this theorem.

Proof: Let us define $G(z, \alpha)$ as the result of operating with $\underset{y}{O}$ on $F_y(z, \alpha)$:

$$G(z, \alpha) \equiv \underset{y}{O}[F_y(z, \alpha)].$$

Then

$$\underset{z}{D}\, G(z, \alpha) = \underset{z}{D}\,\underset{y}{O}[F_y(z, \alpha)],$$

$$= \underset{y}{O}\,\underset{z}{D}[F_y(z, \alpha)],$$

by the first commutability property (ii). Then since the $F_y(z, \alpha)$ are solutions of the F-equation,

$$\underset{z}{D}\, G(z, \alpha) = \underset{y}{O}[F_y(z, \alpha+1)].$$

Similarly,

$$\underset{\alpha}{E}\, G(z, \alpha) = \underset{\alpha}{E}\,\underset{y}{O}[F_y(z, \alpha)],$$

$$= \underset{y}{O}\,\underset{\alpha}{E}[F_y(z, \alpha)],$$

by the second commutability property (ii). Then

$$\underset{\alpha}{E}\, G(z, \alpha) = \underset{y}{O}[F_y(z, \alpha+1)].$$

Hence $G(z, \alpha)$ is a solution of the F-equation. From the third commutability property (ii) it follows that

$$G(z_0, \alpha) = \underset{y}{O}[F_y(z_0, \alpha)].$$

Hence by the formula (1),

$$G(z_0, \alpha) = \phi(\alpha), \qquad R\alpha \geq \alpha_0.$$

Then formula (2) follows from Corollary [11.5]. Q.E.D.

Note: The form of the operator O_y is purposely left perfectly general. In particular, while we conceive its use as a generalization of the idea of linear combination, it need not be a linear operator, for we have nowhere needed to use any such property as

$$O_y[f(y, z, \alpha) + g(y, z, \alpha)] = O_y[f(y, z, \alpha)] + O_y[g(y, z, \alpha)].$$

To show that the operator O_y need not be linear, consider a trivial example. Let O_y be the operator such that

$$O_y[f(y, z, \alpha)] = e^z ,$$

whatever is the function $f(y, z, \alpha)$ and whatever the values of z and α. Then O_y certainly binds the variable y. Since

$$O_y[D_z f(y, z, \alpha)] = e^z , \quad D_z O_y[f(y, z, \alpha)] = e^z ,$$

$$O_y[E_\alpha f(y, z, \alpha)] = e^z , \quad E_\alpha O_y[f(y, z, \alpha)] = e^z ,$$

$$O_y[f(y, z, \alpha)]|_{z = z_0} = e^{z_0}, \quad O_y[f(y, z_0, \alpha)] = e^{z_0},$$

this operator possesses all the commutability properties (ii). Hence all conditions of Theorem [16.1] are satisfied, and in this case the theorem tells us that if a solution of the F-equation is independent of α for some fixed value of z, it must be propertional to e^z; in other words, Theorem [16.1] thus includes a special case of Theorem [10.2] (which is of course also an obvious direct consequence of Corollary [11.5]). The above operator O_y is not linear:

$$\underset{y}{O}[f(y, z, \alpha) + g(y, z, \alpha)] = e^z ,$$

$$\underset{y}{O}[f(y, z, \alpha)] + \underset{y}{O}[g(y, z, \alpha)] = 2e^z .$$

The commutability property (ii) demands that

$$\underset{y}{O}[\lim_{w \to 0} \frac{f(y, z+w, \alpha) - f(y, z, \alpha)}{w}] =$$

$$\lim_{w \to 0} \frac{\underset{y}{O}[f(y, z+w, \alpha)] - \underset{y}{O}[f(y, z, \alpha)]}{w};$$

we may say then that $\underset{y}{O}$ must be linear when applied to functions "infinitely near to each other." It is true that in the succeeding nontrivial examples the special forms we shall choose for $\underset{y}{O}$ will all be linear, but it is possible that there exist nonlinear operators $\underset{y}{O}$ having the desired properties (i) and (ii) which lead to special applications of Theorem [16.1] which are not trivial.

COROLLARY [16.2]. Let $F_1(z, \alpha)$ and $F_2(z, \alpha)$ be two solutions of the F-equation. Suppose there exists an operator $\underset{y}{O}$ as in Theorem [16.1], and a function $G(t)$ such that

(3) $$F_2(z_0, \alpha) = F_1(z_0, \alpha)\, \underset{y}{O}[\{G(y)\}^{\alpha}], \qquad R\alpha \geq \alpha_0 .$$

Let the operator $\underset{y}{O}$ be subject to the further restriction

(iii) $$\underset{y}{O}[h(z, \alpha)\, k(y)] = h(z, \alpha)\, \underset{y}{O}[k(y)],$$

for arbitrary functions $h(z, \alpha)$ and $k(y)$.

Then

$$(4)\quad F_2(z,\alpha) = \underset{y}{O}[\{G(y)\}^{\alpha} F_1((z-z_0)G(t) + z_0,\alpha)], \quad R\alpha \geqq \alpha_0,$$

for all values of z for which the expression on the right has a meaning.

Note: The condition (iii) is a type of linearity condition.

Proof: If $F_1(z, \alpha)$ satisfies the F-equation, so does $\{G(y)\}^{\alpha} F_1((z-z_0)G(y) + z_0, \alpha)$, and when $z = z_0$ this function reduces to $F_1(z_0, \alpha)\ \{G(t)\}^{\alpha}$. Let the functions $F_y(z, \alpha)$ in Theorem [16.1] be given by the definition

$$F_y(z, \alpha) = \{G(y)\}^{\alpha} F_1((z-z_0)G(y) + z_0, \alpha).$$

Then

$$\begin{aligned} \underset{y}{O}[F_y(z_0, \alpha)] &= \underset{y}{O}[\{G(y)\}^{\alpha} F_1(z_0, \alpha)], \\ &= F_1(z_0, \alpha)\ \underset{y}{O}[\{G(y)\}^{\alpha}], \\ &= F_2(z_0, \alpha), \quad R\alpha \geqq \alpha_0, \end{aligned}$$

by the hypotheses (iii) and (3) respectively. Hence all conditions of Theorem [16.1] are satisfied in this special case. Q.E.D.

Example [16.1]. Let us set ourselves the following problem: To find a relation expressing the Laguerre function in terms of the Bessel function $J_a(x)$. We select the solutions nos. 37 and 26 of §5 for consideration. We shall use the method of Corollary [16.2]. In the notation of that corollary, let us write

$$F_1(z, \alpha) \equiv e^{i\alpha\pi} z^{-\frac{\alpha}{2}} J_\alpha(2\sqrt{z}),$$

$$F_2(z, \alpha) \equiv e^{i\alpha\pi} e^{-z} L_c^{(\alpha)}(z).$$

Then

$$F_1(0, \alpha) = \frac{e^{i\alpha\pi}}{\Gamma(\alpha+1)},$$

$$F_2(0, \alpha) = \frac{e^{i\alpha\pi}\Gamma(\alpha+c+1)}{\Gamma(\alpha+1)\Gamma(c+1)}.$$

Then to satisfy equation (3) in Corollary [16.2] we need to find a function $G(y)$ and an operator $\underset{y}{O}$ such that

$$\frac{\Gamma(\alpha+c+1)}{\Gamma(c+1)} = \underset{y}{O}[\{G(y)\}^\alpha].$$

A suitable selection is given by the formulas

$$G(y) \equiv y,$$

$$\underset{y}{O}[\dots] \equiv \frac{1}{\Gamma(c+1)} \int_0^\infty e^{-y} y^c [\dots] dy,$$

as we may see from formula (A.3). Then by the formula (4) of Corollary [16.2],

$$e^{i\alpha\pi} e^{-z} L_c^{(\alpha)}(z) = \frac{e^{i\alpha\pi}}{\Gamma(c+1)} \int_0^\infty e^{-y} y^{c+\alpha} (zy)^{-\frac{\alpha}{2}} J_\alpha(2\sqrt{zy}) dy,$$

or, more conveniently,

$$(5) \qquad e^{-z} z^{\frac{\alpha}{2}} L_c^{(\alpha)}(z) = \frac{1}{\Gamma(c+1)} \int_0^\infty e^{-y} y^{c+\frac{\alpha}{2}} J_\alpha(2\sqrt{zy}) dy.$$

We have thus discovered Hankel's formula (1.33).

Let us remark one similarity and one difference between the preceding derivation of this formula and our previous deduction of it as a special case of formula (15.7). In neither case did we say, Here is a formula, let us use properties of the F-equation to prove it. A formula of this type, once discovered, is usually quite easy to prove; we have used in each case a method of discovery, so that one is led naturally to this formula even if he has no initial idea what it is, beyond in the former case that it is an integral of the type considered in Corollary [15.2] involving the solution

$$\text{(A)} \qquad e^{i\alpha\pi} z^{-\frac{\alpha}{2}} J_\alpha(2\sqrt{z})$$

of the F-equation, and in the latter case that it is a relation giving

$$e^{i\alpha\pi} e^{-z} L_c^{(\alpha)}(z)$$

in terms of the solution (A). In the first derivation we showed how to evaluate a general type of integral whose integrand contained a solution of the F-equation; when this solution is chosen to be (A), in one special case of the result the integral simply happens to turn out a Laguerre function. In the second derivation we have shown how a person wishing specifically to find an expression for the Laguerre function in terms of the Bessel function $J_a(x)$ will be led directly by Corollary [16.2] to the formula (5), without having any initial idea that that relation will be of integral type.

It is natural to ask whether the formula (5) can be inverted. To use Corollary [16.2] we need to interchange the roles of $F_1(z, \alpha)$ and $F_2(z, \alpha)$; that is, we need to find a function $G(y)$ and an operation $\underset{y}{O}$ such that

$$F_1(0, \alpha) = F_2(0, \alpha) \underset{y}{O}[\{G(y)\}^\alpha],$$

or

$$\frac{\Gamma(c+1)}{\Gamma(\alpha+c+1)} = \underset{y}{O}[\{G(y)\}^{\alpha}].$$

We easily see from Hankel's formula (A.1) that a suitable choice is given by the formulas

$$G(y) \equiv \frac{1}{y},$$

$$\underset{y}{O}[\ldots] \equiv \frac{\Gamma(c+1)}{2\pi i} \int_{-\infty}^{(0+)} e^{y} y^{-c-1} [\ldots] dy.$$

Then by Corollary [16.2],

$$(6) \quad z^{-\frac{\alpha}{2}} J_{\alpha}(2\sqrt{z}) = \frac{\Gamma(c+1)}{2\pi i} \int_{-\infty}^{(0+)} e^{y - \frac{z}{y}} y^{-\alpha-c-1} L_c^{(\alpha)}(\tfrac{z}{y}) dy.$$

We have thus discovered the required inversion formula. If we put c equal to zero we recover Schläfli's integral (1.21) for the Bessel function $J_a(x)$.

It is easy to generalize this integral, once it has been suggested to us:

$$(7) \quad \frac{1}{2\pi i} \int_{-\infty}^{(0+)} e^{y - \frac{z}{y}} y^{-\alpha-c} L_b^{(\alpha)}(\tfrac{z}{y}) dy$$

$$= \frac{\Gamma(\alpha+b+1)}{\Gamma(\alpha+1)\Gamma(b+1)\Gamma(\alpha+c)} \, {}_1F_2(\alpha+b+1;\ \alpha+1,\ \alpha+c;\ -z);$$

from this formula the relation (6) follows if $c = b + 1$.

Example [16.2]. We may similarly use Corollary [16.2] to find relation between $J_{\alpha}(x)$ and $J_{\alpha+b}(x)$. For both $F_1(z, \alpha)$ and $F_2(z, \alpha)$ we use solution no. 37 of §5:

$$F_2(z, \alpha) \equiv e^{i(\alpha+b)\pi} z^{-\frac{\alpha+b}{2}} J_{\alpha+b}(2\sqrt{z}),$$

$$F_1(z, \alpha) \equiv e^{i\alpha\pi} z^{-\frac{\alpha}{2}} J_{\alpha}(2\sqrt{z}).$$

Then to satisfy the condition (3) we need to find a function $G(y)$ and an operator $\underset{y}{O}$ such that

$$\frac{e^{ib\pi}}{\Gamma(\alpha+b+1)} = \frac{1}{\Gamma(\alpha+1)} \underset{y}{O}[\{G(y)\}^{\alpha}], \qquad R\alpha \geqq \alpha_0.$$

In view of Wallis's formula (B.3) a suitable choice is given by the expressions

$$G(y) = \sin^2 y,$$
$$\underset{y}{O}[\ldots] = \frac{e^{ib\pi}}{2\Gamma(b)} \int_0^{\frac{1}{2}\pi} \cos^{2b-1} y \sin y [\ldots] dy.$$

It follows from Corollary [16.2] that

$$z^{-\frac{\alpha+b}{2}} J_{\alpha+b}(2\sqrt{z})$$
$$= \frac{1}{2\Gamma(b)} \int_0^{\frac{1}{2}\pi} (z \sin^2 y)^{-\frac{\alpha}{2}} J_c(2\sqrt{z} \sin y) \sin^{2\alpha+1} y \cos^{2b-1} y \, dy$$

or more conveniently,

$$(8) \quad \left(\frac{w}{2}\right)^{-b} J_{\alpha+b}(w) = \frac{1}{2\Gamma(b)} \int_0^{\frac{1}{2}\pi} J_{\alpha}(w \sin y) \sin^{2\alpha+1} y \cos^{2b-1} y \, dy.$$

This result, which is valid when $Rb > \frac{1}{2}$ and $R(\alpha + \frac{1}{2}) > 0$, is due to Sonine [Sonine 1, p. 36; Whittaker and Watson 1, p. 382, ex. 23, p. 383, ex. 31; Copson 1, p. 270, ex. 24]; we have already derived another form of it in formula (15.13), and yet another is the formula (1.30).

Example [16.3]. To find a formula giving the Hermite function in terms of the Legendre function, consider the solutions no's. 33 and 15 of §5:

$$F_1(z, \alpha) \equiv e^{-z^2} H_{\alpha-b}(-z),$$

$$F_2(z, \alpha) \equiv \Gamma(\alpha-b+1)(z^2+1)^{-\frac{\alpha+1}{2}} P_\alpha^b\left(-\frac{z}{\sqrt{z^2+1}}\right).$$

Then from formulas (E.11) and (D.17) we see that

$$F_1(0, \alpha) = \frac{2^{\alpha-b}}{\sqrt{\pi}} \cos \tfrac{1}{2}(\alpha-b)\pi \, \Gamma\left(\frac{\alpha}{2} - \frac{b}{2} + \frac{1}{2}\right),$$

$$F_2(0, \alpha) = \frac{2^\alpha}{\pi} \cos \tfrac{1}{2}(\alpha-b)\pi \, \Gamma\left(\frac{\alpha}{2} + \frac{b}{2} + \frac{1}{2}\right) \Gamma\left(\frac{\alpha}{2} - \frac{b}{2} + \frac{1}{2}\right).$$

The condition (3) then becomes

$$\frac{2^b \Gamma\left(\frac{\alpha}{2} + \frac{b}{2} + \frac{1}{2}\right)}{\sqrt{\pi}} = \underset{y}{O}[\{G(y)\}^\alpha].$$

Hence we may see by Euler's integral (A.3) that a suitable choice for the function G and the operator $\underset{y}{O}$ is given by the definitions

$$G(y) \equiv \sqrt{y},$$

$$\underset{y}{O}[\ldots] \equiv \frac{2^b}{\sqrt{\pi}} \int_0^\infty e^{-y} \, y^{-\frac{1}{2} + \frac{b}{2}} [\ldots] dy.$$

Substitution in the formula (4) yields at once the result

$$\Gamma(\alpha-b+1)(z^2+1)^{-\frac{\alpha+1}{2}} P_\alpha^b\left(\frac{z}{\sqrt{z^2+1}}\right)$$

$$= \frac{2^b}{\pi} \int_0^\infty e^{-y(1+z^2)} \, y^{\frac{1}{2}(\alpha+b-1)} \, H_{\alpha-b}(z\sqrt{y}) dy,$$

which is easily rearranged into the form

$$(9) \qquad (1-x^2)^{\frac{b}{2}} P_\alpha^b(x) = \frac{2^{b+1}}{\sqrt{\pi}\,\Gamma(\alpha-b+1)} \int_0^\infty e^{-t^2} \, t^{\alpha+b} \, H_{\alpha-b}(xt) dt.$$

Our original formula (1.34), the special case of formula (9) which results by putting b equal to zero, was discovered by Glaisher and extended to the case when α is complex by Curzon [Glaisher 1, p. 126; Curzon 1, p. 238; Whittaker and Watson 1, p. 332, ex. 16]. While the formula (9) may be deduced from the Glaisher-Curson integral by differentiation in the case when b is a positive integer, not only is our method of derivation more elementary but also it is valid whenever $R(\alpha + b + 1) > 0$.

By these methods it is equally easy to deduce an inverse for the formula (9):

$$e^{-z^2} H_a(z)$$

$$= \frac{\Gamma(a+1)\sqrt{\pi}}{2^a \cdot 2\pi i} \int_{-\infty}^{(0+)} e^w w^{-\frac{b}{2}} (z^2+w)^{-\frac{b+a+1}{2}} P_{b+a}^{b}\left(\frac{z}{\sqrt{z^2+w}}\right) dw .$$

By an easy transformation of this result we see that

$$(10)\quad H_a(z) = \frac{\Gamma(a+1)}{2^a\sqrt{\pi}\, i} \int_{+\infty}^{-\infty} e^{-y^2} (-y^2-z^2)^{-\frac{b}{2}} (-y)^{-b-a} P_{b+a}^{b}\left(\frac{iz}{y}\right) dy,$$

the contour of integration being indented so as to pass above the origin. The special case of this formula when $b = 0$ was discovered by Curzon [Curzon 1, p. 245].

Example [16.4]. Consider the solutions nos. 26 and 19 of §5:

$$F_1(z, \alpha) \equiv e^{z-1} L_b^{(\alpha)}(1-z),$$

$$F_2(z, \alpha) \equiv (1-z^2)^{-\frac{1}{2}\alpha} P_b^{\alpha}(z).$$

By formulas (E.1) and (D.12) we see that

$$F_1(1, \alpha) = \frac{\Gamma(\alpha+b+1)}{\Gamma(\alpha+1)\Gamma(b+1)} ,$$

$$F_2(1, \alpha) = \frac{\Gamma(+b+1)}{2^{\alpha}\Gamma(b-\alpha+1)\Gamma(\alpha+1)} .$$

By using a process similar to that of the previous examples we easily show that

$$(11)\quad P_b^{\alpha}(z)$$

$$= \frac{(1-z^2)^{\frac{1}{2}\alpha}\,\Gamma(b+1)}{2^{\alpha}\cdot 2\pi i}\int_{-\infty}^{(0+)} e^{\frac{1}{2}y(z+1)}\, y^{\alpha-b+1}\, L_b^{(\alpha)}(\tfrac{1}{2}y[1-z])dy.$$

Example [16.5]. Theorem [16.1] is useful also for summing series. As an example we shall find with its aid a generating function for the Laguerre functions. Let us write

$$A(z) = \sum_{y=0}^{\infty} w^y\, L_y^{(\alpha)}(z) .$$

Then

$$e^{i\alpha\pi}e^{-z}A(z) = \sum_{y=0}^{\infty} w^y e^{i\alpha\pi}e^{-z}\, L_y^{(\alpha)}(z) .$$

Here, in the notation of Theorem [16.1]:

$$\underset{y}{O}[\dots] = \sum_{y=0}^{\infty} w^y[\dots],$$

$$F_y(z,\ \alpha) = e^{i\alpha\pi}e^{-z}\, L_y^{(\alpha)}(z).$$

Hence $e^{i\alpha\pi}e^{-z}A(z)$ is a solution $F(z,\ \alpha)$ of the F-equation, such that

$$F(0,\ \alpha) = \underset{y}{O}[F_y(0,\ \alpha)],$$

$$= \sum_{y=0}^{\infty} w^y e^{i\alpha\pi}\,\frac{\Gamma(\alpha+y+1)}{\Gamma(\alpha+1)y!}\ ,$$

$$= e^{i\alpha\pi}(1-w)^{-\alpha-1}.$$

From Theorem [11,4], or by inspection, we may see that

$$F(z, \alpha) = e^{i\alpha\pi}(1-w)^{-\alpha-1} e^{-\frac{z}{1-w}};$$

then from Theorem [16.1] we have shown that

$$(12) \qquad \sum_{y=0}^{\infty} w^y L_y^{(\alpha)}(z) = (1-w)^{-\alpha-1} e^{-\frac{zw}{1-w}}.$$

One generalization of this formula we have already given, formula (14.4). Theorem [16.1] enables us to find another generalization. Consider the solution no. 10 of §5:

$$F_y(z, \alpha) \equiv e^{i\alpha\pi}\Gamma(\alpha) z^{-\alpha} F(\alpha, -y; b; \tfrac{1}{z}).$$

Then by Gauss's formula (C.4),

$$F_y(1, \alpha) = \frac{e^{i\alpha\pi}\Gamma(\alpha)\Gamma(b)}{\Gamma(b-\alpha)} \cdot \frac{\Gamma(b-\alpha+y)}{\Gamma(b+y)},$$

providing $R(b - \alpha + y) > 0$. Let us define $\underset{y}{O}$ as follows:

$$\underset{y}{O}[\ldots] \equiv \frac{1}{2\pi i} \int_{-\infty i}^{\infty i} \Gamma(-y)(-w)^y [\ldots] dy .$$

Then

$$\underset{y}{O}[F_y(1, \alpha)] = e^{i\alpha\pi}\Gamma(\alpha) \; {}_1F_1(b-\alpha; b; w),$$

from Barnes's integral (C.1). Hence, by Theorems [16.1] and [11.5],

$$(13) \quad \sum_{n=0}^{\infty} \Gamma(\alpha+n) \; {}_1F_1(b-\alpha-n; b; w) \frac{(1-z)^n}{n!}$$

$$= \frac{\Gamma(\alpha) z^{-\alpha}}{2\pi i} \int_{-\infty i}^{\infty i} \Gamma(-y)(-w)^y F(\alpha, -y; b; \tfrac{1}{z}) dy.$$

If in this formula we put b equal to α, it reduces to a formula equivalent to formula (12).

Example [16.6]. Given the problem, To find a formula expressing the confluent hypergeometric function in terms of the ordinary hypergeometric function, whether or not one achieves success and which of the several such formulas one discovers will depend upon which solutions involving these two functions are selected from §5. Let us begin with the choice (nos. 10 and 29):

$$F_2(z, \alpha) \equiv \Gamma(\alpha)(-z)^{-\alpha} F(\alpha, b; c; -\tfrac{1}{z}),$$

$$F_1(z, \alpha) \equiv \Gamma(\alpha)(-z)^{-\alpha} \; {}_1F_1(\alpha; d; \tfrac{w}{z}),$$

where we have transformed solution no. 29 with the aid of the definition (E.1). From the series (C.2) and Gauss's formula (C.4), respectively, we find that

$$F_1(-1, \alpha) = \Gamma(d) \sum_{y=0}^{\infty} \frac{(-w)^y}{y!} \frac{\Gamma(\alpha+y)}{\Gamma(d+y)},$$

$$\cdot F_2(-1, \alpha) = \frac{\Gamma(\alpha)\Gamma(c)\Gamma(c-\alpha-b)}{\Gamma(c-\alpha)\Gamma(c-b)}.$$

The occurrence of $-\alpha$ in the argument of the Γ-functions in $F_2(-1, \alpha)$ suggests that our connection could be established if the $F_1(z, \alpha)$ solution involved $-\alpha$ also. Hence we use Kummer's first transformation (1.17):

$$e^w F_1(-1, \alpha) = \Gamma(\alpha) \; {}_1F_1(d-\alpha; d; w),$$

$$= \frac{\Gamma(\alpha)\Gamma(d)}{\Gamma(d-\alpha)} \sum_{y=0}^{\infty} \frac{\Gamma(d-\alpha+y)}{\Gamma(d+y)} \cdot \frac{w^y}{y!}.$$

We now find exactly $F_2(-1, \alpha)$ occurring in the summands on the right if we make the choice $d = c$, $b = -y$. Then

$$e^w F_1(-1, \alpha) = \sum_{y=0}^{\infty} \frac{w^y}{y!} F_2(-1, \alpha).$$

Hence in the notation of Theorem [16.1],

$$F_y(z, \alpha) = \Gamma(\alpha)(-z)^{-\alpha} F(\alpha, -y; c; -\tfrac{1}{z}),$$

$$\underset{y}{O}[\ldots] = e^{-w} \sum_{y=0}^{\infty} [\ldots] \frac{w^y}{y!} .$$

Then by Theorem [16.1],

$$(14) \qquad e^w \, {}_1F_1(\alpha; c; \tfrac{w}{z}) = \sum_{y=0}^{\infty} \frac{w^y}{y!} F(\alpha, -y; c; -\tfrac{1}{z}) .$$

This formula, which may be considered a generalization of Kummer's first transformation, was discovered by Humbert [Humbert 1, p. 64; Bateman 2, p. 460, ex. 9]; it should be compared with our formula (14.5). The preceding rather clumsy derivation could easily be polished, but it reproduces the actual steps of discovery one goes through, starting with only the problem stated at the beginning of this example and a knowledge of Theorem [16.1] and the two particular solutions of the F-equation in question. It illustrates a slightly more difficult type of discovery than that shown in examples [16.1] through [16.4] because the resulting formula is not included in the rather specific form of Corollary [16.2].

In his original derivation Humbert observed that $F(a, -m; b; y)$ is a polynomial from which a solution of Appell's differential-difference equation may be constructed, and then used Appell's method of finding a generating function for this polynomial. In our language of the F-equation, Humbert substituted a suitably modified form of solution no. 10 of §5 into formula (38). We might have added the derivation of formula (14) to our other examples of the use of Theorem [14.4]. Thus the formula (14) may easily be discovered by the theory of the F-equation in two ways: first as a solution of the problem "Find a generating function for $F(a, -m; b; y)$," which we attack with the aid of Theorem [14.4], and second as a solution of the problem "Find a formula giving

${}_1F_1(a; b; w)$ in terms of $F(a_1, b_1; c; y)$," which we attack with the aid of Theorem [16.1].

COROLLARY [16.3]. If $F_1(z, \alpha)$ and $F_2(z, \alpha)$ are solutions of the F-equation, and if for some one value of z the relation

$$F_2(z, \alpha) = \underset{y}{O}[F_1(z+y, \alpha)], \qquad R\alpha \geq \alpha_0, \tag{15}$$

is correct, then it is correct for all values of z for which the expression on the right has a meaning.

This corollary is useful in generalizing formulas.

Example [16.7]. Let us begin with Sonine's integral (1.28), which we have already proved as a special case of formula (15.9):

$$\int_0^\infty e^{-wy} \frac{y^{-\alpha} L_c^{(-\alpha)}(y)}{\Gamma(-\alpha+c+1)} dy = \frac{(w-1)^c}{\Gamma(c+1)w^{-\alpha+c+1}} . \tag{16}$$

The integral on the left results from putting z equal to zero in the integral

$$I = \int_0^\infty e^{-wy} \frac{(y+z)^{-\alpha} L_c^{(-\alpha)}(y+z)}{\Gamma(-\alpha+c+1)} dy.$$

Now

$$I = \underset{y}{O}[F_1(z+y, \alpha)] ,$$

where

$$\underset{y}{O}[\ldots] = \int_0^\infty e^{-yw}[\ldots]dy$$

$$F_1(z, \alpha) = \frac{z^{-\alpha} L_c^{(-\alpha)}(z)}{\Gamma(-\alpha+c+1)} .$$

$F_1(z, \alpha)$ satisfies the F-equation (v. §5, no. 27). Hence, to evaluate the integral I, by Corollary [16.3] we need only find a solution of the F-equation which reduces when $z = 0$ to

$$\frac{(w-1)^c w^{\alpha-c-1}}{\Gamma(c+1)} \ .$$

By inspection, or by Theorem [11.7], we may see that

$$F(z, \alpha) = \frac{(w-1)^c w^{\alpha-c-1}}{\Gamma(c+1)} \ e^{zw} \ .$$

Hence

$$(17) \qquad \int_0^\infty e^{-yw}(y+z)^\alpha L_c^{(\alpha)}(y+z)dy = \frac{\Gamma(\alpha+c+1)(w-1)^c}{\Gamma(c+1)w^{\alpha+c+1}} \ e^{zw} \ .$$

COROLLARY [16.4]. If $F_1(z, \alpha)$ and $F_2(z, \alpha)$ are solutions of the F-equation, and if for some value of z the relation

$$(18) \qquad F_2(z, \alpha) = \underset{y}{O}[F_2(z, \alpha+y)], \qquad R\alpha \geqslant \alpha_o,$$

is correct, then it is correct for all values of z for which the expression on the right has a meaning.

Example [16.8]. Consider the series

$$F_2(z, \alpha; w) = \sum_{y=0}^\infty e^{i(\alpha+y)\pi} \frac{w^y}{y!} z^{-\frac{\alpha+y}{2}} J_{\alpha+y}(2\sqrt{z}).$$

Clearly $F_2(z, \alpha; w)$ is of the form (18), where

$$\underset{y}{O}[\ldots] \equiv \sum_{y=0}^\infty \frac{w^y}{y!} [\ldots],$$

$$F_1(z, \alpha) \equiv e^{i\alpha\pi} z^{-\frac{\alpha}{2}} J_\alpha(2\sqrt{z}).$$

Now

$$F_2(0, \alpha; w) = \sum_{y=0}^{\infty} \frac{e^{i\alpha\pi}(-)^y w^y}{y!\Gamma(\alpha+y+1)},$$

$$= e^{i\alpha\pi} w^{-\frac{\alpha}{2}} J_\alpha(2\sqrt{w}).$$

Hence, by Theorem [11.5],

$$F_2(z, \alpha; w) = \sum_{y=0}^{\infty} e^{i(\alpha+y)\pi} w^{-\frac{\alpha+y}{2}} J_{\alpha+y}(2\sqrt{w}) \frac{z^y}{y!}.$$

It follows from Corollary [16.4] that

$$(19) \quad \sum_{y=0}^{\infty} \frac{(-)^y w^y}{y!} z^{-\frac{\alpha+y}{2}} J_{\alpha+y}(2\sqrt{z}) = \sum_{y=0}^{\infty} \frac{(-)^y z^y}{y!} w^{-\frac{\alpha+y}{2}} \cdot J_{\alpha+y}(2\sqrt{w}),$$

in other words, $F_2(z, \alpha; w) = F_2(w, \alpha; z)$.

Example [16.9]. As a final example of,the use of Theorem [16.1], we shall show how a little ingenuity will lead to less obvious formulas. It is know that [Milne-Thomson 1, p. 321, ex. 24]

$$(20) \qquad \frac{z^{y-1}}{\Gamma(y)} = \sum_{n=0}^{\infty} (-)^n L_n(z)\binom{y-1}{n}.$$

Hence [Milne-Thomson 1, p. 309]

$$L_n(z) = (-)^n \underset{y}{\Delta}^n \frac{z^{y-1}}{\Gamma(y)}\Big|_{y=1}.$$

Now by Koshliakov's formula (1.32),

$$L_n^{(\alpha)}(z) = \frac{\Gamma(\alpha+n+1)}{\Gamma(\alpha)n!} \int_0^1 (1-t)^{\alpha-1} L_n(zt)dt.$$

Hence

$$L_n^{(\alpha)}(z) = \frac{\Gamma(\alpha+n+1)}{\Gamma(\alpha)n!} \int_0^1 (1-t)^{\alpha-1} (-)^n \Delta_y^n \frac{(zt)^{y-1}}{\Gamma(y)}\Big|_{y=1} dt,$$

$$= \frac{(-)^n\Gamma(\alpha+n+1)}{\Gamma(\alpha)n!} \Delta_y^n\Big\{\frac{z^{y-1}}{\Gamma(y)} \int_0^1 (1-t)^{\alpha-1}t^{y-1}dt\Big\}_{y=1},$$

$$= \frac{(-)^n\Gamma(\alpha+n+1)}{n!} \Delta_y^n \frac{z^{y-1}}{\Gamma(\alpha+y)}\Big|_{y=1}.$$

Then

$$\text{(A)}\quad e^{i\alpha\pi}e^{-z}L_n^{(\alpha)}(z)\Big|_{z=1} = \frac{(-)^n e^{i\alpha\pi}\Gamma(\alpha+n+1)}{e\ n!} \Delta_y^n \frac{1}{\Gamma(\alpha+y)}\Big|_{y=1}.$$

Consider solution no. 26 of §5:

$$F(z, \alpha) \equiv e^{i\alpha\pi}e^{-z}L_n^{(\alpha)}(z).$$

Let us define $h(y, \alpha)$ and $\underset{y}{O}$:

$$h(y, \alpha) \equiv \frac{e^{i\alpha\pi}\Gamma(\alpha+n+1)}{\Gamma(\alpha+y)},$$

$$\underset{y}{O}[\ldots] \equiv \frac{(-)^n}{e\ n!} \Delta_y^n [\ldots]_{y=1}.$$

Then by formula (A),

$$F(1, \alpha) = \underset{y}{O}[h(y, \alpha)].$$

Hence, by Theorems [11.5] and [16.1],

$$F(z, \alpha) = \frac{(-)^n e^{i\alpha\pi}}{e\ n!} \Delta_y^n\Big\{ \sum_{m=0}^{\infty} \frac{(-)^m\Gamma(\alpha+n+m+1)}{\Gamma(\alpha+m+y)} \cdot \frac{(z-1)^m}{m!}\Big\}_{y=1}.$$

Then

$$L_n^{(\alpha)}(z) = \frac{(-)^n e^{z-1}\Gamma(\alpha+n+1)}{n!} \Delta_y^n \frac{{}_1F_1(\alpha+n+1;\ \alpha+y;\ 1-z)}{\Gamma(\alpha+y)}\Big|_{y=1}.$$

By Kummer's first transformation (1.17), we then show that

$$L_n^{(\alpha)}(z) = \frac{(-)^n\Gamma(\alpha+n+1)}{n!} \underset{y}{\Delta}^n \frac{{}_1F_1(\alpha-n-1;\ \alpha+y;\ z-1)}{\Gamma(\alpha+y)}\Bigg|_{y=1},$$

or, in terms of Laguerre polynomials,

$$L_n^{(\alpha)}(z+1) = \frac{(-)^n}{n!} \underset{y}{\Delta}^n \{\Gamma(n+2-y)L_{n+1-y}^{(\alpha+y-1)}(z)\}_{y=1}.$$

By the operator identity $\Delta = E - 1$,

$$L_n^{(\alpha)}(z+1) = \frac{(-)^n}{n!} \sum_{m=0}^{n} (-)^m\binom{n}{m}[\Gamma(m+2-y)L_{m+1-y}^{(\alpha+y+n-m-1)}(z)]_{y=1}.$$

We have then discovered the addition formula

$$(21) \qquad L_n^{(\alpha)}(z+1) = \sum_{m=0}^{n} \frac{(-)^m}{m!} L_{n-m}^{(\alpha+m)}(z).$$

Chapter V

REMARKS ON SOLUTIONS SUCH THAT $F(z, \alpha_0) = \psi(z)$

§17. THE UNIQUENESS OF SOLUTIONS SUCH THAT $F(z, \alpha_0) = \psi(z)$

It is evident that if the function $F(z, \alpha)$ satisfies the equation (3.4) and the condition

$$(1) \qquad F(z, \alpha_0) = \psi(z),$$

then, if $\psi(z)$ is n times differentiable the functions $F(z, \alpha_0 + 1)$, $F(z, \alpha_0 + 2)$, ..., $F(z, \alpha_0 + n)$ may be uniquely calculated from the functional equation. If $\psi(z)$ is arbitrarily many times differentiable, than a unique solution $F(z, \alpha)$ exists for all values of α which exceed α_0 by an integer. For values of α_0 which are less than α by a positive integer m, a solution $F(z, \alpha_0)$ exists, provided $\psi(z)$ is integrable, but that solution contains m arbitrary constants.

Suppose we have a function $F(z, \alpha)$ satisfying the equation (3.4) which is defined for all values of α in some continuous range, and which satisfies the condition (1). Suppose that $F_1(z, \alpha)$ is <u>any</u> solution of the equation (3.4), defined for the same range of α, and that $\pi(\alpha)$ is any periodic function of α of period 1 such that $\pi(\alpha_0) = 0$. Then the function

$$F(z, \alpha) + \pi(\alpha) F_1(z, \alpha)$$

is also a solution of the equation (3.4) and also satisfies the condition (1). It is then clear that there is no simple uniqueness theorem concerning solutions satisfying the boundary condition (1).

Any formulas whose deduction is based on comparing the values of two solutions of the F-equation when α is fixed at α_0 are then valid necessarily only for values of α which exceed α_0 by a positive integer. In the succeeding paragraphs we use the boundary condition $F(z, \alpha_0) = \psi(z)$, first in the discovery and proof of certain formulas which are valid only when $\alpha = \alpha_0 + n$, and second in the discovery and conjecture of certain formulas valid more generally.

§18. DIFFERENTIAL FORMULAS

In this section we answer Question 3 of §1.

The most immediately obvious property of solutions of the F-equation is expressed by the following theorem:

THEOREM [18.1]. If $F(z, \alpha)$ satisfies the F-equation, then

$$F(z, \alpha+n) = \frac{\partial^n}{\partial z^n} F(z, \alpha). \tag{1}$$

Example [18.1]. Substitution of solutions no. 37 and 33 of §5 in the formula (1) leads at once to formulas (1.11) and (1.13), respectively, if $\alpha = 0$. If we substitute solution no. 15 of §5 into the formula (1), put b equal to α, and use the formula (D.18), we obtain after some rearrangement the formula

$$P^{\alpha}_{\alpha+n}(\cos\theta) = \frac{(-)^n \Gamma(2\alpha+1)}{n!\; 2^{\alpha}\Gamma(\alpha+1)} \csc^{\alpha+n+1}\theta \, \frac{d^n \sin^{2\alpha+1}\theta}{d\,(\cot\theta)^n}, \tag{2}$$

of which the known formula (1.16) is a special case [Whittaker and Watson 1, p. 331, ex. 15]. With the solution no. 25 of §5 we may proceed in exactly the same fashion, except that not knowing an analogue of the formula (D.18) to evaluate $Q^{\alpha}_{\alpha}(t)$ in terms of simple functions, we are forced to consider the special case in which $\alpha = 0$. With the aid of formula (D.25) we may then prove that

$$(3)\quad Q_n(\cos\theta) = \frac{(-)^n}{n!} \csc^{n+1}\theta \frac{d^n}{d(\cot\theta)^n} \{\sin\theta \log(\cot\theta + \csc\theta)\}.$$

I have seen neither the formula (2) nor the formula (3) in the literature; the latter must be known, however, since the late Professor Bateman set the discovery of an analogue for the function $Q_n(x)$ to the formula (1.16) as an examination question in his course in Potential Theory in May, 1942, but none of the students solved the problem and he did not disclose the formula he had in mind. Putting the solution no. 24 of §5 when $\alpha = 0$ into our formula (1), we discover the formula

$$(4)\quad P_n^{(b,-b)}(\cos\theta) = \frac{(-)^n}{n!} \csc^{n+1}\theta(\csc\theta+\cot\theta)^b \frac{d^n}{d(\cot\theta)^n}\{\sin\theta(\cot\theta+\csc\theta)^b\}.$$

Using the solutions no. 28 and no. 26 of §5, respectively, we may derive from formula (1) the formulas

$$(5)\qquad L_n^{(c)}(y) = \frac{(-)^n}{n!} y^{-1-c} e^y \frac{d^n}{d(\frac{1}{y})^n} [y^{1+c} e^{-y}] ,$$

$$(6)\qquad L_c^{(n)}(y) = (-)^n e^y \frac{d^n}{dy^n} e^{-y} L_c(y).$$

If we already have the formula (1.14), then Deruyts' formula (1.15) [Deruyts 1, p. 9; Bateman 2, p. 452; Copson 1, p. 269, ex. 23] is an obvious consequence of the formula (6).

We see that the formulas (1.11), (1.13), a generalization of formula (1.16), and formula (1.15) are all immediate special cases of the formula (1). The proper analogue of these formulas for the Laguerre functions considered as functions of their argument and lower index is not the formula (1.14) but the formula (5).

Now it is true that the formula (1.14) is a consequence of the formula (5), and that Rodrigues' formula (1.12) can be proved from the formula (1.16), but the transformations are not obvious and do not seem to be general. Our analysis suggests that possibly there may be a second type of differential formula involving n^{th} powers. While it is true that the expression

$$(7) \quad F(z, \alpha+n) = \frac{\partial^n}{\partial h^n} \left[F(z(h), \alpha)\, z'(h) \left\{ \frac{h - h_0}{z(h) - z_0} \right\}^n \right]_{h=h_0}$$

where $h = h(z)$, $z = z(h)$, $h_0 = h(z_0)$, is a generalization of the formula (1), it is not the desired generalization of Rodrigues' formula,first because it does not tell us what function $h(z)$ to choose, and second because it is not quite of the same form. I suspect that the Rodrigues' formula (1.12) and the formula (1.14) are not direct and natural consequences of the fact that transforms of the functions involved are solutions of the F-equation; as reasons I suggest first that the Jacobi polynomial possesses a Rodrigues' formula even though it does not possess a transform which satisfies the F-equation, while the analogue of the formula (1) for the equation (3.4), viz.

$$(8) \quad F(y, \alpha+n)$$

$$= \frac{1}{C(y, \alpha+n-1)} \frac{\partial}{\partial y} \left[\frac{1}{C(y, \alpha+n-2)} \frac{\partial}{\partial y} \cdots \frac{1}{C(y, \alpha)} \frac{\partial}{\partial y} F(y, \alpha) \right],$$

is sufficiently complicated that it is unlikely to be formally useful, and second that the analogue of Rodrigues' formula for the function $Q_n(x)$ involves not differentiation but integration [Whittaker and Watson 1, p. 318, ex. 2], while the recurrent properties of the functions $P_n(x)$ and $Q_n(x)$ are identical. In a sense, of course, we may say that we have already deduced the Rodrigues' formula (1.12) and the formula (1.14) from the theory of the F-equation, since they are special cases of the contour integrals (1.22) and (1.24), whose natural discovery we have sketched and achieved respectively in Examples [13.5] and [13.3], but the interpretation of these results as n^{th} derivative formulas seems to be accidental and not characteristic of the general contour integrals discussed in §11.

COROLLARY [18.2]. If $F(z, \alpha)$ is a solution of the F-equation and if $F(z, \alpha_0)$ is expressible as a Laplace transform,

$$F(z, \alpha_0) = \int_0^\infty e^{-tz} h(t)dt, \tag{9}$$

then so also is $F(z, \alpha_0 + n)$:

$$F(z, \alpha_0 + n) = (-)^n \int_0^\infty e^{-tz} t^n h(t)dt. \tag{10}$$

Example [18.2]. Consider solution no. 15 of §5 when $\alpha = 0$, $b = 0$. Then

$$F(z, 0) = (z^2 - 1)^{-\frac{1}{2}},$$

$$= \int_0^\infty e^{-zt} I_0(t)dt, \qquad |z| > 1.$$

[Churchill 1, p. 297, no. 49.] Hence by formula (10),

$$n!\ (z^2 - 1)^{-\frac{n+1}{2}} P_n\left(- \frac{z}{\sqrt{z^2-1}}\right) \tag{11}$$

$$= (-)^n \int_0^\infty e^{-tz} t^n I_0(t)dt, \qquad |z| > 1.$$

This result we used previously in Example [14.16]; it is of course a special case of our formula (15.7).

Example [18.3]. From the definition (F.6) and a simple modification of solution no. 41 of §5 we may see that $F(z, \alpha)$ as given by the definition

$$F(z, \alpha) \equiv e^{i\alpha\pi} z^{-\frac{\alpha}{2}} K_{-\alpha}(2\sqrt{z})$$

is a solution of the F-equation. Then

$$F(z, 0) = K_0(2\sqrt{z}),$$
$$= \int_0^\infty e^{-zt - \frac{1}{t}} \frac{dt}{t}$$

[Churchill 1, p. 301, no. 117]. Hence by formula (10),

$$z^{-\frac{n}{2}} K_{-n}(2\sqrt{z}) = \frac{1}{2} \int_0^\infty e^{-zt - \frac{1}{t}} t^{n-1}\, dt. \tag{12}$$

This formula can be shown to be valid for all values of n, real or complex [Magnus and Oberhettinger 1, p. 128]. It is apparently equivalent to a formula due to Hobson [Watson 1, p. 172].

Example [18.4]. Consider solution no. 28 of §5:

$$F(z, \alpha) = e^{i\alpha\pi} \Gamma(\alpha+1) z^{-\alpha-1-b} e^{-\frac{1}{z}} L_\alpha^{(b)}(z).$$

Then

$$F(z, 0) = z^{-1-b} e^{-\frac{1}{z}},$$
$$= \int_0^\infty e^{-zt} t^{\frac{b}{2}} J_b(2\sqrt{t})dt, \qquad Rb > -1,$$

[Churchill 1, p. 299, no. 80]. By application of the formula (10) we deduce the integral

$$n!\, z^{-n-1-b} e^{-\frac{1}{z}} L_n^{(b)}(\tfrac{1}{z}) = \int_0^\infty e^{-zt} t^{n + \frac{b}{2}} J_b(2\sqrt{t})dt, \tag{13}$$

a special case of our earlier formula (15.7).

The preceding examples indicate that the use of Laplace transforms in dealing with solutions of the F-equation is both insufficiently general and insufficiently suggestive. The results we deduced are only special cases of more general results obtainable with our integrals of §15, which we there discovered in a much more natural fashion, employing much less information about the functions involved. Theorem [18.1] becomes more useful when we apply it in combination with Theorem [16.1], as the following example shows:

Example [18.5]. Consider solution no. 28 of §5,

$$F_y(z, \alpha) \equiv \Gamma(\alpha+1)(-\tfrac{1}{z})^{\alpha+y+1} e^{-\frac{1}{z}} L_\alpha^{(y)}(\tfrac{1}{z}).$$

Suppose $|z| > 1$. Then if

$$\underset{y}{O}[\ldots] \equiv \int_0^\infty y^{b-1}[\ldots]dy$$

it follows that $\underset{y}{O}[F_y(z, \alpha)]$ as given by the formula

$$\underset{y}{O}[F_y(z, \alpha)] = \Gamma(\alpha+1)e^{-\frac{1}{z}} \int_0^\infty y^{b-1} e^{-(\alpha+y+1)\log(-z)} L_\alpha^{(y)}(\tfrac{1}{z})dy$$

is a solution of the F-equation. Now

$$\underset{y}{O}[F_y(z, 0)] = e^{-\frac{1}{z}}(-\tfrac{1}{z}) \int_0^\infty y^{b-1} e^{-y\log(-z)} dy,$$

$$= -\tfrac{1}{z} e^{-\frac{1}{z}} [\log(-z)]^b \Gamma(b).$$

Then by Theorem [18.1] and Theorem [16.1],

$$\underset{y}{O}[F_y(z, n)] = \Gamma(b) \frac{d^n}{dz^n}\{-\tfrac{1}{z} e^{-\frac{1}{z}} [\log(-z)]^b\}.$$

In other words

$$(14) \quad \int_0^\infty t^{b-1} w^{n+t+1} L_n^{(t)}(-w)dt$$

$$= \frac{(-)^n \Gamma(b)}{n!} e^{-w} \frac{d^n}{d(\frac{1}{w})^n} \{we^w [\log \frac{1}{w}]^b\}, \qquad Rb > 0, \ |w| < 1 .$$

§19. INTEGRAL FORMULAS

Since by Theorem [18.1] and Cauchy's formula it follows that

$$(1) \qquad F(z, \alpha+n) = \frac{n!}{2\pi i} \int_C \frac{F(y, \alpha)}{(y-z)^{n+1}} dy,$$

where C is a contour surrounding the point $y = z$, the following theorem is suggested:

> THEOREM [19.1]. If $F(z, \alpha)$ is a solution of the F-equation and an analytic function of z in a domain including the point z and in a region containing the line extending from the point z toward $-\infty$ and parallel to the imaginary axis, then

$$(2) \qquad F(z, \alpha+n) = \frac{\Gamma(n+1)}{2\pi i} \int_{-\infty}^{(0+)} y^{-n-1} F(z+y, \alpha)dy.$$

Our theorem is stated unnecessarily weakly because in this form it suggests a formula which may be valid for non-integral values of n.

Example [19.1]. If we apply Theorem [19.1] to solution no. 45 of §5, observing that

$$F(z, 0) = \frac{e^z}{1 - e^z},$$

we obtain the integral

$$(3)\quad \Phi(w, -\alpha) = \frac{\Gamma(\alpha+1)}{2\pi i} \int_{-\infty}^{(0+)} \frac{y^{-\alpha-1}}{e^{-y}-w}\, dy, \qquad \alpha = 0, 1, 2, \ldots .$$

By other means it is easy to show that this integral is valid for all except negative integer values of α. An equivalent result has been given by Jonquière [Jonquière 1; Truesdell 1, p. 146].

In order to answer Question 7 of §1 we must consider the case when n is a negative integer; the formula (2) becomes indeterminate, but the following theorem can be proved:

THEOREM [19.2]. Suppose $F(z, \alpha)$ is any solution of the F-equation such that

(i) $F(y, \alpha_0)$ is an integrable function of t in the range $z_0 \leqq y \leqq z$:

(ii) $F(z_0, \alpha_0 - p)$, $p = 0, 1, \ldots, m$ exists;

Then $F(z, \alpha_0 - m)$ exists and is uniquely defined by the formula

$$(4)\quad F(z, \alpha_0 - m)$$

$$= \int_{z_0}^{z} F(y, \alpha_0) \frac{(z-y)^{m-1}}{(m-1)!}\, dy + \sum_{p=1}^{m} F(z_0, \alpha_0 - p) \frac{(z-z_0)^{m-p}}{(m-p)!} .$$

Proof: Since

$$\frac{\partial}{\partial z} F(z, \alpha_0 - 1) = F(z, \alpha_0)$$

it follows that

$$F(z, \alpha_0 - 1) = \int_{z_0}^{z} F(y_1, \alpha_0)\, dy_1 + F(z_0, \alpha_0 - 1).$$

Similarly

$$F(z, \alpha_0 - 2)$$

$$= \int_{z_0}^{z} F(y_2, \alpha_0 - 1)dy_2 + F(z_0, \alpha_0 - 2),$$

$$= \int_{z_0}^{z}\int_{z_0}^{y_2} F(y_1, \alpha_0)dy_1 dy_2 + (z - z_0)F(z_0, \alpha_0 - 1) + F(z_0, \alpha_0 - 2).$$

By integrating m times we show that

$$F(z, \alpha_0 - m)$$

$$= \int_{z_0}^{z}\int_{z_0}^{y_m}\cdots\int_{z_0}^{y_2} F(y_1, \alpha_0)dy_1 dy_2 \ldots dy_m + \sum_{p=1}^{m} F(z_0, \alpha_0 - p)\frac{(z - z_0)^{m-p}}{(m-p)!}$$

The formula (4) now follows from elementary calculus.

The formula (4) is due to Doetsch, who discovered and proved it in a more elaborate fashion [Doetsch 1, p. 262], and showed that it will give the formula (11.24) for the Charlier polynomials in the case when both α and b are integers. It is clear, in fact, that when $F(z, \alpha_0) = 0$, this formula gives the coefficients of the polynomial $F(z, \alpha_0 - m)$ in terms of $F(z_0, \alpha_0 - m)$, where z_0 is any number whatever.

Example [19.2]. The solution no. 34 of §5 becomes infinite when $\alpha = -p$, but by the choices

$$b = -1, \ c = 0, \ \pi(\alpha) = \frac{e^{-i\alpha\pi} \sin \alpha\pi}{\pi}$$

in Corollary [7.3] we may deduce from solution no. 34 another which is more suitable for our present purpose:

$$F(z. \alpha) = \frac{H_{-\alpha}(\frac{1}{2} z)}{\Gamma(1-\alpha)} .$$

Then

$$F(z, 1) = 0,$$

and from formula (E.11) we see that

$$F(z, 1-p) = \frac{2^{-1+p}}{\sqrt{\pi}} \sin\frac{p\pi}{2} \cdot \frac{\Gamma(\frac{p}{2})}{\Gamma(p)} .$$

Substitution in formula (4) then yields the formula

$$(5) \qquad H_m(\tfrac{1}{2} z) = \frac{m!}{\sqrt{\pi}} \sum_{n=0}^{m} 2^n \cos\frac{n\pi}{2} \frac{\Gamma(\frac{n}{2}+\frac{1}{2})}{n!(m-n)!} z^{m-n} ,$$

equivalent to the formula (E.10) when $a = m$.

In this way we may easily find expressions for Legendre and Laguerre polynomials also.

Much more interesting is the use of the formula (4) to deduce integro-difference relations, as illustrated by the succeeding examples.

Example [19.3]. If we substitute solution no. 37 of §5 into the formula (4) we have the result

$$z^{-\frac{\alpha_0-m}{2}} J_{\alpha_0-m}(2\sqrt{z}) = \frac{1}{(m-1)!} \int_z^{z_0} (t-z)^{m-1} t^{-\frac{\alpha_0}{2}} J_{\alpha_0}(2\sqrt{t})dt$$

$$+ \sum_{j=1}^{m} z_0^{-\frac{\alpha_0-j}{2}} J_{\alpha_0-j}(2\sqrt{z_0}) \frac{(z_0-z)^{m-j}}{(m-j)!} ,$$

$$= \frac{1}{(m-1)!} \int_0^{z_0-z} t^{m-1}(z+t)^{-\frac{\alpha_0}{2}} J_{\alpha_0}(2\sqrt{z+t})dt$$

$$+ \sum_{j=1}^{m} z_0^{-\frac{\alpha_0-j}{2}} J_{\alpha_0-j}(2\sqrt{z_0}) \frac{(z_0-z)^{m-j}}{(m-j)!} .$$

When $1 \leqq m < R(\alpha_0/2 + 1/4)$ we are justified in letting z_0 approach ∞, obtaining the formula

$$(6) \quad z^{-\frac{\alpha-m}{2}} J_{\alpha-m}(2\sqrt{z}) = \frac{1}{\Gamma(m)} \int_0^{\infty} t^{m-1}(z+t)^{-\frac{\alpha}{2}} J_{\alpha}(2\sqrt{z+t})dt ,$$

a special case of Sonine's formula (1.31) [Sonine 1, p. 38; Copson 1, p. 343, ex. 14].

This sort of use of formula (4) should be regarded as a method of discovery only. Once we see the formula (6) before us, it is natural to ask if it remains true when m is not an integer. That it does is easy to show in a variety of ways. For example, we need only notice that when $1 \leq Rm < R(\alpha_0/2 + 1/4)$ the integral remains convergent, and that if we multiply through by $e^{i(\alpha_0 - m)\pi}$ each side becomes a solution of the F-equation. When $z = 0$, the left side obviously reduces to $e^{i(\alpha_0-m)\pi}/\Gamma(\alpha_0-m+1)$, and it is a consequence of Weber's formula (1.26) that the right side does also. The generalized equation (6) then follows from Corollary [11.3].

Thus our Theorem [19.2] does not give us a sufficiently general proof of the formula (6), but it gives a speedy and efficient method of discovery.

It is worth mention that all our procedure is equally effective if we replace the function $J_\alpha(x)$ by the function $Y_\alpha(x)$.

Example [19.4]. By substituting solution no. 38 of §5 into the formula (4) we have the result

$$z^{\frac{\alpha+m}{2}} J_{\alpha+m}(2\sqrt{z}) = \frac{1}{\Gamma(m)} \int_{z_0}^{z} y^{\frac{\alpha}{2}}(z-y)^{m-1} J_\alpha(2\sqrt{y})dy$$

$$+ \sum_{p=1}^{m} z_0^{\frac{\alpha+p}{2}} J_{\alpha+p}(2\sqrt{z_0}) \frac{(z-z_0)^{m-p}}{(m-p)!} .$$

If $R\alpha > -1$ and m is any positive integer we may put z_0 equal to zero and find that

$$(7) \qquad z^{\frac{\alpha+m}{2}} J_{\alpha+m}(2\sqrt{z}) = \frac{1}{\Gamma(m)} \int_{0}^{z} y^{\frac{\alpha}{2}}(z-y)^{m-1} J_\alpha(2\sqrt{y})dy$$

a special case of Sonine's formula (1.30), which we have

already discovered and proved twice for unrestricted values of m as our formulas (15.13) and (16.8).

Example [19.5]. Substituting solution no. 27 of §5 into the formula (4), we have at once the result

$$z^{\alpha+m}L_b^{(\alpha+m)}(z) = \frac{\Gamma(\alpha+m+b+1)}{\Gamma(m)} \int_{z_0}^{z} y^{\alpha}(z-y)^{m-1}L_b^{(\alpha)}(y)dy$$

$$+ \sum_{p=1}^{m} \binom{\alpha+m+b+1}{m-p} z_0^{\alpha+j}L_b^{(\alpha+p)}(z_0)(z-z_0)^{m-p}.$$

If $R\alpha > -1$ we may put z_0 equal to zero and obtain the formula

$$(8)\quad z^{\alpha+m}L_b^{(\alpha+m)}(z) = \frac{\Gamma(\alpha+m+b+1)}{\Gamma(m)} \int_{0}^{z} y^{\alpha}(z-y)^{m-1}L_b^{(\alpha)}(y)dy ,$$

a special case of Koshliakov's formula (1.32), which we have already generalized by our formula (15.15).

Example [19.6]. If we substitute the solution no. 15 of §5 in the formula (4) we obtain the result

$$(9)\quad \Gamma(\alpha-b-m+1)(z^2-1)^{-\frac{\alpha-m+1}{2}} P_{\alpha-m}^{b}\left(-\frac{z}{\sqrt{z^2-1}}\right)$$

$$= \int_{z_0}^{z} \Gamma(\alpha-b+1)(y^2-1)^{-\frac{\alpha+1}{2}} P_{\alpha}^{b}\left(-\frac{y}{\sqrt{y^2-1}}\right) \frac{(z-y)^{m-1}}{(m-1)!}\, dy$$

$$+ \sum_{p=1}^{m} \frac{\Gamma(\alpha-b-p+1)}{(m-p)!} (z-z_0)^{m-p} (z_0^2-1)^{-\frac{\alpha-p+1}{2}} P_{\alpha-p}^{b}\left(-\frac{z_0}{\sqrt{z_0^2-1}}\right).$$

If $m < R\alpha$ and $b + p - \alpha \neq$ a positive integer for $p = 1, 2, \ldots, m$, the series vanishes as $z_0 \to \infty$. Simplifying the result by an obvious change of variables, we may then deduce the formula

$$(10) \quad (x^2-1)^{\frac{\alpha}{2}} P^b_{\alpha-m}(x)$$

$$= m\binom{\alpha-b}{m} \int_{-1}^{x} (t^2-1)^{\frac{\alpha-m-1}{2}} (t\sqrt{x^2-1}-x\sqrt{t^2-1})^{m-1} P^b_\alpha(t)dt \ .$$

From these examples we observe that the same expansion formula (4) gives at once the explicit form of polynomial solutions of the F-equation, and a class of integro-difference relations including our original formulas (1.30), (1.31), and (1.32).

It is interesting to contrast our two derivations of the formulas (1.30) and (1.32). In Examples [15.5] and [15.6] they were both found to be very special cases of the general integral formula (15.10); we had no initial idea what the results would be, but we knew we could evaluate the integrals involved in terms of simple series, providing we used transforms (in this case no. 37 and no. 26 of §5 respectively) whose values were known explicitly for some fixed value of z. Our results were valid for all values of α in a right half plane. In the derivation we have just given we used the formula (4), which tells us at once that $F(z, \alpha_0-m)$ is expressible as the sum of an integral involving $F(z, \alpha_0)$ plus a finite sum which vanishes in these cases. Our results are valid only for restricted values of α, they do not appear as special cases of more general formulas involving the same functions, but they are very directly and immediately obtained, and we were able to use in their discovery transforms (no. 38 and no. 27 of §4) which we do not know as functions of α in any simple explicit fashion for any fixed value of z.

Chapter VI

CONCLUSIONS AND UNSOLVED PROBLEMS

§20. HISTORICAL NOTE

Some special functional equations reducible to the F-equation were studied in the 18th century by Taylor, the Bernoullis, Euler, and other classical mathematicians (cf. equation (9.3)). For some notion of the known methods of treating particular equations of this type the reader is referred to a recent paper of Bateman where six types of attack, of which we have made some use of the first five in this essay, are outlined [Bateman 3]. So far as I know, however, there has not been presented prior to the present work what might be called a systematic treatment of these equations as a class, and with a view to unifying the theory of familiar special functions.

Lommel seems to have been the first to realize that a differential-difference equation could be used to characterize functions satisfying it, even for values of the differenced variable not congruent mod 1 [Lommel 2]. He observed that the equations

$$(1) \qquad \frac{\partial}{\partial z} z^{\nu} J_{\nu}(z) = z^{\nu} J_{\nu-1}(z) ,$$

$$(2) \qquad z^{\nu} J_{\nu}(z) = 2(\nu-1) z^{\nu-1} J_{\nu-1}(z) - z^{\nu} J_{\nu-2}(z)$$

were sufficient to characterize the Bessel function $J_{\nu}(x)$, but he did not make much use of the fact. From the point of view of this essay we may say that he adds to the F-equation (in a particular case) a difference equation rather than a boundary condition to secure uniqueness of a particular solution.

The only attempt, so far as I know, at a systematic treatment of an equation equivalent to the F-equation in which the differenced variable is not restricted to values congruent mod 1 is contained in the opening portion of a great memoir of Sonine, to which we have often referred in the foregoing pages [Sonine 1]. Sonine begins by considering the equation

$$S_{n+1}(x) + 2\frac{dS_n(x)}{dx} - S_{n-1}(x) = 0, \tag{3}$$

n being supposed a positive integer and $S_0(x)$ an arbitrary function. He discovers the expansion

$$S_0(\alpha+x) = J_0(\alpha)S_0(x) + 2\sum_{n=1}^{\infty} (-)^n J_n(\alpha)S_n(x), \tag{4}$$

generalizing a result of C. Neumann. Supposing further that

$$n\,S_n(x) = \frac{x}{2}[S_{n-1}(x) + S_{n+1}(x)] , \tag{5}$$

he finds four contour integrals which satisfy both the equations (3) and (5), for any value of n, which by a complicated argument he shows to represent Bessel functions. In the first part of his memoir Sonine eschews any use of Bessel's differential equation; he shows that any solution of the system (3), (5) which can be represented by an integral of the form

$$\int_a^b \Phi(t, x)t^{-n-1}dt \tag{6}$$

must differ only trivially from $J_n(x)$ or $Y_n(x)$, deduces the values a, b, and $\Phi(t, x)$ are permitted to have, and upon the resulting formulas builds his further investigations.

Sonine's method is to progress from formula to formula, using one to beget another, until he accumulates an impressive list of relations involving Bessel functions.

After the opening pages of his memoir, he does not return to equation (3). He makes no study of equations equivalent to the equation (3), nor does he apply his results to functions other than Bessel functions. In fact, by the addition of the equation (5), he rules out results applicable to other functions. His memoir is a study of Bessel functions which by a slight change of approach might have become a study of a much larger class of functions.

In the same year as Sonine's memoir appeared also a note by Appell [Appell 1] on polynomials $A_n(x)$ satisfying the equation

$$(7) \qquad A_n'(x) = n\, A_{n-1}(x).$$

Appell deduced a generating expansion equivalent to that of our Theorem [14.4], part III, and several interesting consequences of that formula. He remarked that the equation

$$(8) \qquad X_n'(x) = \phi(n)\, X_{n-1}(x)$$

is reducible to the equation (7), but he did not mention that much more general equations are so reducible and that his results may be applied to many familiar functions. He considered only the case when n is a positive integer. Forty-four years later Humbert pointed out the power of Appell's method in a paper [Humbert 1] in which he applied it to various types of hypergeometric polynomials in one and and two variables. In particular, he used it to rediscover Sonine's formula (1.8) and to discover his formula (16.14) and several more complicated series involving his Sonine polynomials in two variables. That Appell's generating expansion is not very well known is indicated by our Examples [14.10] to [14.14], which deduce from it as very easy special cases several formulas usually proved by more cumbersome methods and not previously known to be in any way related to each other.

Bruwier in a recent paper considers the equation

$$(9) \qquad F(x, n; y_n, y_n', \ldots, y_n^{(p)}; y_{n+1}, \ldots, y_{n+q}) = 0$$

where n is a positive integer [Bruwier 1]. He begins by remarking that the most general integral of the equation [ibid., p. 10]

$$(10) \qquad y_n' = a\, y_{n+1} + F_n(x)$$

is

$$(11) \qquad y_n = \frac{D^{n-1} y_1}{a^{n-1}} - \frac{D^{n-2}}{a^{n-2}} F_1 - \cdots - \frac{F_{n-1}}{a},$$

or, if $F_n(x)$ is not differentiable,

$$(12) \qquad y_n = \frac{D^{n-1} y_1}{a^{n-1}} + \int_{x_0}^{x} \sum_{k=0}^{\infty} a^k \frac{(x-z)^k}{k!} F_{n+k}(z)dz.$$

The particular integral may be written symbolically as [ibid., pp. 23-24]

$$(13) \qquad y_n = \frac{I}{1 - aEI} F_n,$$

where I is the integral operator. Similarly a particular integral for the equation

$$(14) \qquad y_n^{(p)} = a\, y_{n+q} + F_n(x)$$

is

$$(15) \qquad y_n = \frac{I^p}{1 - aE^q I^p} F_n,$$

$$= \int_{x_0}^{x} \sum_{k=0}^{\infty} \frac{a^k (x-z)^{kp+p-1}}{(kp+p-1)!} F_{n+kq}(z)dz.$$

Bruwier then shows that the equation [ibid., pp. 33-36]

$$(16) \qquad y_n'(x) + A(x)y_n(x) = B(x)y_{n+1}(x) + F_n(x)$$

is reducible to the equation

$$(17) \qquad y_n' = B(x)y_{n+1} + F_n(x) ,$$

which in its turn is reducible to the form

$$(18) \qquad y_n' = y_{n+1} + F_n(x) ,$$

and hence its general solution may be written down as above. He similarly obtains a symbolic formula for a particular integral of the equation

$$(19) \quad y_n^{(p)} + A_1(x)y_n^{(p-1)} + \dots + A_p(x)y_n$$

$$= B_1(x)y_{n+1} + \dots + B_q(x)y_{n+q} + F_n(x).$$

At the end of the paper [ibid., pp. 43-48] Bruwier proves an existence theorem for the equation

$$(20) \qquad y_n' = F_n(x, y_n, y_{n+1})$$

with the condition $y_n(x_o) = y_o$, y_o being a function of n, from which he sketches a formulation and proof of a similar theorem for the equation (9). This result is the equivalent of our Theorem [9.1].

Throughout his investigation Bruwier apparently restricts himself to the case when n is an integer, and any explicit solutions he exhibits are not valid for nonintegral values of n. He does not apply his results to familiar special functions.

In his study of the Charlier polynomials Doetsch has considered the equation [Doetsch 1]

$$\frac{\partial}{\partial t}\Phi(x, t) = \frac{\Phi(x-h, t) - \Phi(x, t)}{h}. \tag{21}$$

He discovers our series (11.14) and our expansion (19.4), the latter with the quite unnecessary aid of the convolution integral for Laplace transforms. He does not mention the fact that other functional equations are reducible to equation (21), nor does he apply his results to functions other than the Charlier polynomials and a simple generalization of them.

We may summarize these results from the more general outlook of this essay:

1. Sonine and Doetsch both considered the case when (in the notation of the F-equation) α is not necessarily an integer, but they confined themselves to the cases of Bessel and Charlier functions, respectively, and Sonine added such further restrictions that his results were not applicable to other functions.

2. Bruwier proved a theorem equivalent to the most useful result of this essay, but he proved it (apparently) for integer values of α only, and did not apply it to special functions of interest.

3. Appell deduced an expansion theorem which we have found as a special case of one of ours, but did not apply it to special functions of interest other than the Hermite polynomials. Humbert indicated the usefulness of Appell's expansion by giving several interesting special cases.

4. Appell and Bruwier each made different preliminary steps toward the essential reduction we have given in §3; but neither of these writers noticed that these reductions might be applied to the recurrence relations satisfied by familiar special functions.

I laid out the groundwork and found the main results of this essay before becoming acquainted with any of these memoirs, simply following what seemed the most logical, direct, and thorough course of investigation.

§21. SOME CONCLUSIONS FROM THIS ESSAY

We have seen that many relations satisfied by familiar special functions are special cases of several simple theorems concerning solutions of the F-equation. I have refrained from laboring the analytic side of these investigations at the expense of the formal, since in general the special formulas I have given as examples are fairly easily proved, once set up, and if proved in a restricted fashion, may easily be generalized. The real questions of interest are:

1. How do we discover such formulas?

2. Are they deducible from any sort of more general relation, or are they the peculiar property of the special function involved?

Towards answering these two questions, which are closely connected, this essay has been directed. We have shown that many special functions can easily be transformed into solutions of the F-equation by simple changes of variable, and that the resulting transforms satisfy numerous interesting formal relations in virtue of their being such solutions.

We have exhibited many such relations explicitly, and showed by examples that they included as special cases numerous familiar and unfamiliar formulas scattered through the literature of special functions and discovered in different ways by different persons, and other formulas besides, which, to the best of my knowledge, were previously unknown.

We have shown fruitful methods of discovery for many other formulas, enabling one to attack with good promise of success even a problem as vague as "given two special solutions $F_1(z, \alpha)$ and $F_2(z, \alpha)$ of the F-equation, find a formula giving $F_2(z, \alpha)$ in terms of $F_1(z, \alpha)$, knowing that $F_2(z_0, \alpha) = \phi_2(\alpha)$, $F_1(z_1, \alpha) = \phi_1(\alpha)$."

The methods of this essay are so easy to apply that I believe an investigator wishing a formula of one of the types discussed here would be able to rediscover it in less time than a search of the literature might require.

In general the examples I have given require but scanty knowledge of the functions involved; I have avoided more recondite ones because I wished to emphasize the ease with which apparently complicated results could be discovered and proved by these methods.

All the results of this essay are not only analytically but even formally very elementary. I have included only results that flow easily and directly from the F-equation, showing that many formulas which previous investigators required pages of space and sometimes complicated apparatus to deduce are in fact elementary or trivial if considered from the point of view of the F-equation.

It would of course be possible to go from formula to formula, deriving more and more complicated results, and still using only the F-equation fundamentally. A finished treatment of special functions, however, should rely exclusively on no one approach; were a new treatise to be written, I believe it should employ those tools which lend themselves to the most efficient and elegant deductions, and that one of those tools would be the theory of the F-equation.

§22. SOME UNANSWERED QUESTIONS

Naturally there are many interesting formulas which we have not been able to subsume under our analysis. While since each special function is different from each other one, each must certainly have <u>some</u> properties which cannot follow from some common property of this and other functions, there are a number of relations I have observed in the literature which I feel should follow directly from theorems on solutions of the F-equation, but which I have not been able to generalize properly. As each successive revision of this essay has been completed, the list has shrunk, but there still remain a few items.

I. We have given no discussion of singular solutions, that is, functions satisfying the F-equation and having, for some fixed value z_0 of z, some sort of singularity as functions of α. An example is the solution no. 38 of §5,

$$F(z, \alpha) \equiv e^{i\alpha\pi} \Gamma(\alpha) z^{-\alpha},$$

since

$$\lim_{|z| \to 0} F(z, \alpha) = \begin{cases} \infty, & R\alpha > 0, \\ 0, & R\alpha < 0, \end{cases} \quad \alpha \text{ fixed},$$

and there are numerous other examples on the list of §5. Results involving these singular solutions are particularly interesting. We have already mentioned the Lindelöf-Wirtinger expansion (9.7); another example is given by the function

(1) $F(z, \alpha)$

$$\equiv \sum_{n=1}^{\infty} d(n) e^{nz} n^{\alpha} - \lim_{k \to 0+} \sum_{n=0}^{\infty} e^{-k(\alpha+n)^2} [\zeta(-\alpha-n)]^2 \frac{z^n}{n!},$$

where $d(n)$ is the divisor function and $\zeta(w)$ is Riemann's zeta-function. It is easy to see that $F(z, \alpha)$ is a solution of the F-equation such that $F(0, \alpha) = 0$ if $R\alpha < -1$, and it would be interesting to establish the value of $F(z, \alpha)$ in general.

There are in fact several solutions listed in §5 whose behavior as functions of α when $z = z_0$ is known in a left half of the α plane, but not in a right half. There are indications that the theory of these functions is more complicated than that we have discussed in the preceding pages most of which applies essentially to functions $F(z, \alpha)$ such that $F(z_0, \alpha)$ is well behaved in a right half plane. A more convenient way of stating this remaining general problem is:

Given a function $H(z, \alpha)$ satisfying the equation

$$\frac{\partial}{\partial z} H(z, \alpha) = H(z, \alpha - 1), \tag{2}$$

to study its properties if the function $H(z_0, \alpha)$ is known and well behaved in a right half plane.

Among solutions of the equation (2) are Dirichlet series of the form

$$\sum_n a_n \frac{e^{nz}}{n^\alpha}, \quad \sum_n \frac{a_n}{n^\alpha} \sin(nz - \tfrac{1}{2}\alpha\pi), \tag{3}$$

where the summation does not necessarily run over integer values of n. Examples of series of the first type occur in the Lindelöf-Wirtinger expansion (9.7) and the first term of the formula (1) above, while an example of the second type is Hurwitz's trigonometric series for the generalized zeta-function [Whittaker and Watson 1, p. 269]. A proper uniqueness theorem for the equation (2) would enable us to show, for example, that the Hurwitz formula is a consequence rather than a generalization of Riemann's functional equation for the zeta-function.

There are also solutions of the F-equation like $e^{i\alpha\pi} z^{-1/2\alpha} Y_\alpha(2\sqrt{z})$ whose singularity at the origin is logarithmic, to which relatively few of our general theorems can conveniently be applied.

II. Among various generating expansions of interest are those of the type

$$\sum a_n F(z, n) F(w, n), \tag{4}$$

where $F(z, \alpha)$ is a solution of the F-equqtion. The form of two familiar expansions of this type [Geronimus 1: Hardy 2; Watson 2; Copson 1, p. 270, ex. 30; p. 271, ex. 34; Szegö 1, p. 98]:

$$(5)\quad \sum_{n=0}^{\infty} \frac{t^n H_n(x) H_n(y)}{2^n n!} = (1-t^2)^{-\frac{1}{2}} \exp\left(\frac{2xyt - (x^2+y^2)t^2}{1-t^2}\right),$$

$$(6)\quad \sum_{n=0}^{\infty} \frac{t^n n! L_n^{(\alpha)}(x) L_n^{(\alpha)}(y)}{\Gamma(\alpha+n+1)}$$

$$= \frac{(xyt)^{-\frac{1}{2}\alpha}}{1-t} \exp\left(-\frac{(x+y)t}{1-t}\right) I_\alpha\left(\frac{2\sqrt{xyt}}{1-t}\right),$$

suggests that the summation of such series in general could be accomplished by means of the F-equation, but so far I have been unable to achieve it. Of course, like all such formulas those above are easy to prove once they are set up. A general sum formula for the series (4) may be written down in terms of either the series (11.5) or the Laplace transforms (18.10), but this type of formula is not sufficiently simple to be really useful:

$$(7)\quad \sum_{n=0}^{\infty} a_n F(w, n) F(y, n)$$

$$= \sum_{m,p=0}^{\infty} \left\{ \sum_{n=0}^{\infty} a_n \phi(n+m) \phi(n+p) \right\} \frac{(w-w_0)^m (y-y_0)^p}{m!\, p!},$$

$$(8)\quad \sum_{n=0}^{\infty} a_n F(w, n) F(y, n)$$

$$= \int_0^{\infty} \int_0^{\infty} e^{-wx-yt} h(x) h(t) \left\{ \sum_{n=0}^{\infty} a_n (xt)^n \right\} dx\, dt.$$

These formulas employ the notation of Theorem [11.2] and Corollary [18.2], respectively. When some special values are given to the a_n, there remain at least two sums or integrals, respectively, to be evaluated in succession.

None of the general formulas given in this essay employ a succession of integrals or series. When I say I have not been able to evaluate the sum (4) I mean that I have been unable to discover a simple expression for it involving a known function, a single integral, or a single simpler sum.

III. Consideration of asymptotic formulas I have reserved for the subject of a subsequent paper.

Appendix I

SPECIAL FUNCTIONS

In this appendix we give the definitions of particular special functions which we have used in this essay, along with those of their formal properties which we have considered given. When the definition set down is valid only in a restricted region, the values of the function elsewhere are obtained by analytic continuation. If a defining formula becomes indeterminate for certain values of the variables, the limiting form of the expression is to be used as the definition for these values.

For a more complete list of formulas, see Magnus and Oberhettinger 1.

A. Gamma Function [Euler, Legendre, Gauss, Weierstrass; Hankel 1; Whittaker and Watson 1, pp. 235-264; Copson 1, pp. 205-232; Milne-Thomson 1, pp. 241-260].

(1) $$\frac{1}{\Gamma(y)} = \frac{1}{2\pi i}\int_{-\infty}^{(0+)} e^w w^{-y} dw. \quad \text{(Hankel)}$$

(2) $$\Gamma(y) = \frac{1}{2i \sin \pi y}\int_{-\infty}^{(0+)} e^w w^{y-1} dw, \quad y \neq 0, \pm 1, \pm 2, \ldots. \text{(Hankel)}$$

(3) $$\Gamma(y) = \int_0^\infty e^{-t} t^{y-1} dt, \qquad Ry > 0. \quad \text{(Euler)}$$

(4) $$\Gamma(y+1) = y\,\Gamma(y). \quad \text{(Euler)}$$

(5) $$\Gamma(y)\,\Gamma(1-y) = \pi \csc \pi y. \quad \text{(Euler)}$$

(6) $\Gamma(y)\,\Gamma(y+\frac{1}{n})\,\Gamma(y+\frac{2}{n})\,\ldots\,\Gamma(y+\frac{n-1}{n})$

$$= n^{\frac{1}{2}-ny}\,(2\pi)^{\frac{1}{2}(n-1)}\,\Gamma(ny).$$ (Legendre, Gauss)

B. Beta Function [Euler, Legendre, Pochhammer; Whittaker and Watson 1, pp. 253-264].

(1) $$B(a,\ b) = \frac{-\,e^{-\pi i(a+b)}}{4\sin a\pi\sin b\pi}\int_A^{(1+,0+,1-,0-)} w^{a-1}(1-w)^{b-1}dw.$$

(Jordan, Pochhammer)

(2) $$B(a,\ b) = \frac{\Gamma(a)\,\Gamma(b)}{\Gamma(a+b)}.$$ (Euler)

(3) $$B(a,\ b) = 2\int_0^{\pi/2}\cos^{2a-1}t\,\sin^{2b-1}t\ dt.$$ (Wallis, Euler)

C. Hypergeometric Functions [Euler, Gauss, Kummer; Barnes 1; Whittaker and Watson 1, pp. 281-301; Copson 1, pp. 245-271; Bailey 1].

(1) $${}_pF_n(a_1,a_2,\ldots,a_p;b_1,b_2,\ldots,b_n;y) = \frac{\Gamma(b_1)\Gamma(b_2)\ldots\Gamma(b_n)}{\Gamma(a_1)\Gamma(a_2)\ldots\Gamma(a_p)}$$

$$\cdot\ \frac{1}{2\pi i}\int_{-i\infty}^{+i\infty}\frac{\Gamma(a_1+w)\Gamma(a_2+w)\ldots\Gamma(a_p+w)}{\Gamma(b_1+w)\Gamma(b_2+w)\ldots\Gamma(b_n+w)}\,\Gamma(-w)(-y)^w dw,$$

when none of $a_1,\ a_2,\ \ldots,\ a_p$ are negative integers, $|\arg y| < \pi,\ |y| < 1$. (Barnes)

$${}_pF_n(-m,a_1,\ldots,a_p;b_1,b_2,\ldots,b_n;y) = (-)^m m!$$

$$\frac{\Gamma(b_1)\Gamma(b_2)\ldots\Gamma(b_n)}{\Gamma(a_2)\ldots\Gamma(a_n)}\sum_{j=0}^{m}\frac{\Gamma(a_2+j)\Gamma(a_3+j)\ldots\Gamma(a_p+j)}{\Gamma(b_1+j)\Gamma(b_2+j)\ldots\Gamma(b_n+j)}\,\frac{y^j}{j!\,\Gamma(1+m-j)}.$$

(2) $\ _pF_n(a_1, a_2, \ldots, a_p; b_1, b_2, \ldots, b_n; y)$

$$= \frac{\Gamma(b_1)\Gamma(b_2)\ldots\Gamma(b_n)}{\Gamma(a_1)\Gamma(a_2)\ldots\Gamma(a_p)} \sum_{j=0}^{\infty} \frac{\Gamma(a_1+j)\Gamma(a_2+j)\ldots\Gamma(a_p+j)}{\Gamma(b_1+j)\Gamma(b_2+j)\ldots\Gamma(b_n+j)} \cdot \frac{y^j}{j!},$$

$$p - 1 < n, \quad \text{or} \quad p - 1 = n \quad \text{and} \quad |y| < 1.$$

(3) $$F(a, b; c; y) \equiv {}_2F_1(a, b; c; y).$$

(4) $$F(a, b; c; 1) = \frac{\Gamma(c)\Gamma(c-a-b)}{\Gamma(c-a)\Gamma(c-b)}, \quad R(c-a-b) > 0. \quad \text{(Gauss)}$$

(5) $$F(2a, 2b; a + b + \tfrac{1}{2}; \tfrac{1}{2}) = \frac{\Gamma(a + b + \frac{1}{2})\Gamma(\frac{1}{2})}{\Gamma(a + \frac{1}{2})\Gamma(b + \frac{1}{2})}. \quad \text{(Kummer)}$$

(6) $$(1 - y)^{-a} = F(a, b; b; y).$$

(7) $$\log \frac{1}{1-y} = y\, F(1, 1; 2; y).$$

(8) $$\frac{d}{dy}\, {}_pF_n(a_1, a_2, \ldots, a_p; b_1, b_2, \ldots, b_n; y) = \frac{a_1 a_2 \ldots a_p}{b_1 b_2 \ldots b_n}$$

$$ {}_pF_n(a_1+1, a_2+1, \ldots, a_p+1; b_1+1, b_2+1, \ldots, b_n+1; y).$$

(9) $$\frac{d}{dy}[y^a\, F(a, b; c; y] = y^{a-1}\, a\, F(a+1, b; c; y).$$

(10) $$\frac{d}{dy}[y^{c-1}\, F(a, b; c; y)] = (c-1)y^{c-2}\, F(a, b; c-1; y).$$

(11) $$\frac{d}{dy}[y^{c-a}(1-y)^{a+b-c}\, F(a, b; c; y)]$$

$$= (c-a)y^{c-a-1}(1-y)^{a+b-c-1}\, F(a-1, b; c; y).$$

(12) $\frac{d}{dy}[(1-y)^{a+b-c}\ F(a,\ b;\ c;\ y)]$

$$= \frac{(c-a)(c-b)}{c}\ (1-y)^{a+b-c-1}\ F(a,\ b;\ c+1;\ y).$$

(13) $\frac{d}{dy}[(1-y)^{a}F(a,b;c;y)] = -\frac{a(c-b)}{c}(1-y)^{a-1}F(a+1,b;c+1;y).$

(14) $\frac{d}{dy}[y^{c-1}(1-y)^{b-c+1}F(a,b;c;y)]$

$$= (c-1)y^{c-2}(1-y)^{b-c}\ F(a-1,b;c-1;y).$$

(15) $\frac{d}{dy}[y^{c-1}(1-y)^{a+b-c}F(a,b;c;y)]$

$$= (c-1)y^{c-2}(1-y)^{a+b-c-1}\ F(a-1,b-1;c-1;y).$$

(16) $F(a,\ b;\ c;\ y) = (1-y)^{c-a-b}\ F(c-a,\ c-b;\ c;\ y).$

(This formula follows from a two-fold application of Euler's transformation (1.20).)

D. <u>Legendre</u>, <u>Jacobi</u>, <u>Gegenbauer</u>, <u>and</u> <u>Bateman</u> <u>Functions</u> (Legendre, Laplace, Jacobi, Heine, Ferrers, Gegenbauer, Hobson, Bateman; Whittaker and Watson 1, pp. 302-336; Copson 1, pp. 272-312; Hobson 2; Szegö 1, pp. 57-95; Bateman 1).

(1) $P_a^{(b,c)}(x) = \binom{a+b}{a}\ F(-a,\ a+b+c+1;\ b+1;\ \frac{1}{2} - \frac{1}{2}x).$

(2) $(2a+b+c)(1-x^2)\ \frac{d}{dx}\ P_a^{(b,c)}(x)$

$$= -\ a\{(2a+b+c)x + c - a\}\ P_a^{(b,c)}(x) + 2(a+b)(a+c)\ P_{a-1}^{(b,c)}(x).$$

(3) $(2a+b+c+2)(1-x^2)\ \frac{d}{dx}\ P_a^{(b,c)}(x) = (a+b+c+1)$

$$\{(2a+b+c+2)x+b-c\}\ P_a^{(b,c)}(x) - 2(a+1)(a+b+c+1)\ P_{a+1}^{(b,c)}(x).$$

(4) $$\Theta_a^{b,c}(x) = \frac{(1+x)^{\frac{b}{2}}(1-x)^{\frac{c}{2}}}{2^{\frac{1}{2}(b+c)}\,\Gamma(c+1)} \cdot \frac{\Gamma(\frac{a+b+c}{2}+1)}{\Gamma(\frac{a-b-c}{2}+1)}$$

$$\cdot\ F(-\frac{a-b-c}{2}, \frac{a+b+c}{2}+1; c+1; \frac{1}{2}-\frac{1}{2}x) .$$

(5) $$\Theta_a^{b,c}(x) = \frac{(1+x)^{\frac{b}{2}}(1-x)^{\frac{c}{2}}}{2^{\frac{1}{2}(b+c)}} \cdot \frac{\Gamma(\frac{a+b+c}{2}+1)}{\Gamma(\frac{a-b+c}{2}+1)} P^{(c,b)}_{\frac{1}{2}(a-b-c)}(x).$$

(6) $$(x^2-1)\frac{d}{dx}\Theta_a^{b,c}(x)$$

$$= \frac{1}{2a}\{a^2x+c^2-b^2\}\Theta_a^{b,c}(x) - \frac{(a+b+c)(a+b-c)}{2a}\Theta_{a-2}^{b,c}(x). \quad \text{(Bateman)}$$

(7) $$(x^2-1)\frac{d}{dx}\Theta_a^{b,c}(x) = \frac{(a-b-c+2)(a+c-b+2)}{2(a+2)}\Theta_{a+2}^{b,c}(x)$$

$$-\frac{1}{2(a+2)}\{(a+2)^2x + c^2 - b^2\}\ \Theta_a^{b,c}(x). \quad \text{(Bateman)}$$

(8) $$C_b^a(x) \equiv \binom{b+2a-1}{b} F(-b, b+2a; a+\frac{1}{2}; \frac{1}{2}-\frac{1}{2}x).$$

(9) $$C_b^a(x) = \frac{\Gamma(a+\frac{1}{2})\,\Gamma(b+2a)}{\Gamma(2a)\,\Gamma(b+a+\frac{1}{2})} P_b^{(a-\frac{1}{2}, a-\frac{1}{2})}(x).$$

(10) $$x\frac{d}{dx}C_b^a(x) = 2aC_b^{a+1}(x) - (b+2a)\,C_b^a(x). \quad \text{(Gegenbauer)}$$

We define the Legendre functions, like the Jacobi, Bateman, and Gegenbauer functions, only for values of the argument x such that $-1 \leq x \leq +1$. Definitions which are valid for complex arguments may be found in the references quoted at the beginning of this section. We restrict ourselves to real arguments because the formulas satisfied by the generalized Legendre functions are slightly different from those given here below. A simple adjustment of a real argument formula usually yields the corresponding complex argument formula. See in particular the list of relations given by Magnus and Oberhettinger 1.

(11) $$P_b^a(x) \equiv \frac{1}{\Gamma(1-a)}\left(\frac{1+x}{1-x}\right)^{\frac{a}{2}} F(-b,\ b+1;\ 1-a;\ \tfrac{1}{2} - \tfrac{1}{2}x).$$

(12) $$P_b^a(x) = \frac{(1-x^2)^{\frac{1}{2}a}\,\Gamma(a+b+1)}{2^a\Gamma(a+1)\Gamma(b-a+1)} F(a+b+1, a-b; a+1; \tfrac{1}{2} - \tfrac{1}{2}x)\ .$$

(13) $$P_b^a(x) = \circledR_{2b}^{a,a}(x).$$

(This formula replaces the incorrect formula (5) of Bateman 1, p. 112.)

(14) $$P_a^b(x) = \frac{(1-x^2)^{\frac{1}{2}b}}{2^b} \cdot \frac{\Gamma(a+b+1)}{\Gamma(a+1)} P_{a-b}^{(b,b)}(x).$$

(15) $$C_b^a(x) = \frac{\sqrt{\pi}}{2^{a-\frac{1}{2}}\Gamma(a)} (1-x^2)^{-\frac{1}{2}(a-\frac{1}{2})} P_{a+b-\frac{1}{2}}^{a-\frac{1}{2}}(x).$$

(16) $$P_b^a(x) = \frac{\Gamma(2a+1)}{2^a\Gamma(a+1)} (1-x^2)^{\frac{1}{2}a} C_{b-a}^{a+\frac{1}{2}}(x).$$

(17) $$P_b^a(o) = \frac{\Gamma(a+b+1)}{2^a\Gamma(b-a+1)} \cdot \frac{\cos\frac{b-a}{2}\pi}{\sqrt{\pi}} \cdot \frac{\Gamma(\frac{b}{2} - \frac{a}{2} + \frac{1}{2})}{\Gamma(\frac{b}{2} + \frac{a}{2} + 1)}\ .$$

(18) $$P_b^b(x) = \frac{\Gamma(2b+1)}{2^b\,\Gamma(b+1)}\,(1-x^2)^{\frac{1}{2}b}.$$

(19) $$P_b^{-b}(x) = \frac{1}{2^b\,\Gamma(b+1)}\,(1-x^2)^{\frac{1}{2}b}.$$

(20) $$Q_b^a(x) \equiv \frac{\pi}{2\sin a\pi}\left[P_b^a(x)\cos a\pi - \frac{\Gamma(a+b+1)}{\Gamma(b-a+1)}\,P_b^{-a}(x)\right],$$

$a + b \neq$ a negative integer.

(21) $$Q_b^a(o) = -2^{b-1}\sqrt{\pi}\,\sin\frac{a+b}{2}\pi\,\frac{\Gamma(\frac{a+b+1}{2})}{\Gamma(\frac{b-a+1}{2})}.$$

(22) $$P_a(x) \equiv P_a^o(x).$$

(23) $$Q_a(x) \equiv Q_a^o(x).$$

(24) $$P_{-b-1}^a(x) = P_b^a(x).$$

(25) $$Q_o(x) = \frac{1}{2}\log\frac{1+x}{1-x}.$$

E. Laguerre, Whittaker, Poisson-Charlier, Incomplete Gamma, Weber, and Hermite Functions. [Whittaker and Watson 1, pp. 337-354; Copson 1, pp. 260-271; Szegö 1, pp. 33-34, pp. 96-106; Doetsch 1.]

(1) $$L_c^{(a)}(y) \equiv \binom{a+c}{a}\,{}_1F_1(-c;\,a+1;\,y).$$

(2) $$L_c^{(a)}(o) = \binom{a+c}{a}.$$

(3) $$L_c(y) \equiv L_c^{(o)}(y).$$

(4) $W_{a,b}(y) \equiv -\frac{1}{2\pi i}\Gamma(a+\frac{1}{2}-b)e^{-\frac{1}{2}by}y^{a}$

$$\int_{\infty}^{(0+)}(-w)^{-a-\frac{1}{2}+b}(1+\frac{w}{y})^{a-\frac{1}{2}+b}e^{-w}\,dw,$$

$a-\frac{1}{2}-b \neq$ a negative integer.

(5) $W_{a,b}(y) = \frac{e^{-\frac{1}{2}y}y^{a}}{\Gamma(\frac{1}{2}-a+b)}\int_{0}^{\infty}t^{-a-\frac{1}{2}+b}(1+\frac{t}{y})^{a-\frac{1}{2}+b}e^{-t}dt,$

$a-\frac{1}{2}-b =$ a negative integer.

(6) $W_{a,b}(y) = \frac{\Gamma(-2b)y^{\frac{1}{2}+b}e^{-\frac{1}{2}y}}{\Gamma(\frac{1}{2}-a-b)}\,{}_1F_1(\frac{1}{2}+b-a;\ 2b+1;\ y)$

$$+\frac{\Gamma(2b)y^{\frac{1}{2}-b}e^{-\frac{1}{2}y}}{\Gamma(\frac{1}{2}+b-a)}\,{}_1F_1(\frac{1}{2}-b-a;\ -2b+1;\ y)\ .$$

(7) $y\frac{d}{dy}W_{a,b}(y) = (a-\frac{1}{2}y)\,W_{a,b}(y) - \{b^2-(a-\frac{1}{2})^2\}W_{a-1,b}(y).$

(8) $$D_a(y) \equiv 2^{\frac{1}{2}a+\frac{1}{4}}y^{-\frac{1}{2}}W_{\frac{1}{2}a+\frac{1}{4},\ -\frac{1}{4}}(\frac{1}{2}y^2).$$

(9) $$H_a(y) \equiv 2^{\frac{a}{2}}e^{\frac{1}{2}y^2}D_a(\sqrt{2}y).$$

(10) $H_a(y)$

$$= \frac{2^{a}\,\Gamma(\frac{1}{2})}{\Gamma(\frac{1}{2}-\frac{1}{2}a)}\,{}_1F_1(-\frac{1}{2}a;-\frac{1}{2};y^2) + \frac{2^{a}\Gamma(-\frac{1}{2})y}{\Gamma(-\frac{1}{2}a)}\,{}_1F_1(\frac{1}{2}-\frac{1}{2}a;\frac{3}{2};y^2).$$

(11) $$H_a(o) = \frac{2^a}{\sqrt{\pi}} \cos \frac{a\pi}{2} \Gamma(\frac{a}{2} + \frac{1}{2}).$$

(12) $$p_n(m,a) \equiv \sum_{i=0}^{n} (-)^{n-i}\binom{n}{i} i!a^{-i}\binom{m}{i}.$$ (Doetsch)

(13) $$\psi_o(m,a) = \frac{a^m}{m!}e^{-a}, \ \psi_n(m,a) \equiv p_n(m,a)\psi_o(m,a).$$ (Doetsch)

(14) $$\psi_n(m,o) = (-)^{n-m}\binom{n}{m}.$$ (Doetsch)

(15) $$\psi_n(m,a) = \psi_{n-1}(m-1,a) - \psi_{n-1}(m,a).$$ (Doetsch)

(16) $$\psi_n(m,a) = \frac{\partial^n}{\partial a^n} \psi_o(m,a).$$ (Doetsch)

(17) $$\gamma(a;y) \equiv \Gamma(a) - y^{\frac{1}{2}(a-1)} e^{-\frac{1}{2}a} W_{\frac{1}{2}(a-1),\frac{1}{2}a}(y).$$

(18) $$\frac{d}{dy} \gamma(a,y) = y^{a-1} e^{-y}.$$

(19) $$\frac{d}{dy}\ {}_1F_1(a;\ 2a;\ y)$$

$$= \frac{1}{2}\ {}_1F_1(a;\ 2a;\ y) + \frac{y}{4(2a+1)}\ {}_1F_1(a+1;\ 2a+2;\ y).$$

F. <u>Bessel and Hankel Functions</u>. [Bessel, Neumann, Lommel, Hankel, Sonine; Whittaker and Watson 1, pp. 355-385; Watson 1; Copson 1, pp. 313-344.]

(1) $$J_a(y) \equiv (\tfrac{1}{2}y)^a \sum_{m=0}^{\infty} \frac{(-)^m (\frac{1}{2}y)^{2m}}{m!\ \Gamma(a+m+1)}.$$

(2) $$Y_a(y) \equiv \frac{\cos a\pi\ J_a(y) - J_{-a}(y)}{\sin a\pi}.$$

(3) $$H_a^{(1)}(y) = \frac{1}{\pi i}\int_{-\infty}^{\infty+\pi i} e^{y \operatorname{sh} w - aw} dw, \quad |\arg y| < \tfrac{1}{2}\pi.$$

(4) $$H_a^{(2)}(y) = -\frac{1}{\pi i}\int_{-\infty}^{\infty-\pi i} e^{y \operatorname{sh} w - aw} dw, |\arg y| < \tfrac{1}{2}\pi.$$

(5) $$I_a(y) = i^{-a} J_a(iy).$$

(6) $$K_a(y) = \tfrac{1}{2}\pi\{I_{-a}(y) - I_a(y)\} \cot a\pi.$$

(7) $$J_n(y) = (-)^n J_{-n}(y).$$

G. Miscellaneous Functions. Spence's transcendent: [Truesdell 1]

(1) $$\phi(y,a) = \sum_{n=1}^{\infty} y^n n^{-a}, \quad |y| < 1.$$

ϕ-polynomials: [Milne-Thomson 1, pp. 124-153]
$\phi_n^{(a)}(y)$ is defined by the expansion

(2) $$f_a(w)e^{yw+g(w)} = \sum_{m=0}^{\infty} \frac{w^m}{m!}\,\phi_m^{(a)}(y),$$

where $f_a(w)$ and $g(w)$ are given functions.

(3) $$\frac{d}{dy}\,\phi_n^{(a)}(y) = n\,\phi_{n-1}^{(a)}(y). \quad \text{(Milne-Thomson)}$$

To obtain the Bernoulli polynomials $B_m^{(n)}(y)$, specialize the preceding definition as follows:

(4) $$f_n(w) = t^n(e^t-1)^{-n}, \quad g(w) = 0.$$

(5) $$\frac{d}{dy}\,B_m^{(n)}(y) = \frac{n-m}{n-y}\,B_m^{(n)}(y) - \frac{n}{n-y}\,B_m^{(n+1)}(y).$$

Polygamma function:

$$\Psi_n(y) = \frac{d^n}{dy^n} \log \Gamma(y+1). \tag{6}$$

Generalized zeta function [Hurwitz; Whittaker and Watson 1, pp. 265-280]

$$\zeta(a,y) = -\frac{\Gamma(1-a)}{2\pi i} \int_{\infty}^{(0+)} \frac{(-w)^{a-1} e^{-yw}}{1-e^{-w}} dw. \tag{7}$$

$$\zeta(a,y) = \sum_{n=0}^{\infty} \frac{1}{(y+n)^a}, \quad Ra \geqq 1 + \delta. \tag{8}$$

Appendix II

OPERATORS

We here list the definitions of differential and difference operators, and those of their formal properties we have actually used in the preceding pages.

$$\mathbf{D}_y f(y, a) \equiv \frac{\partial}{\partial y} f(y, a).$$

$$\Delta_a f(y, a) \equiv f(y, a+1) - f(y, a).$$

$$\mathbf{E}_a f(y, a) \equiv f(y, a+1).$$

$$\mathbf{M}_a f(y, a) \equiv \frac{1}{2}[f(y, a) + f(y, a+1)].$$

$$\mathbf{E}_a = 1 + \Delta_a .$$

$$\mathop{\mathbf{S}}_{a_0}^{a} h(v)\Delta v \equiv \lim_{k \to 0+} \{ \int_{a_0}^{\infty} h(v)e^{-kc(v)}dv - \sum_{m=0}^{\infty} h(a+m)e^{-kc(a+m)} \},$$

subject to analytic restrictions as set forth in Nörlund 1, pp. 47-52, 69-94, or Milne-Thomson 1, pp. 209-213, 220-237.

$$\Delta_a \mathop{\mathbf{S}}_{a_0}^{a} h(v)dv = h(a).$$

The class of operators represented by the symbol $\mathbf{O}_y$ is defined in the statement of Theorem [16.1].

Appendix III

EXAMPLES OF EQUATIONS OF TYPE (3.4) NOT REDUCIBLE TO THE F-EQUATION

I know only five examples of functions which satisfy an equation of type (3.4) which is not reducible to the F-equation. The functions and coefficients $C(y, \alpha)$ in equation (3.4) are listed below.

(1) $g(y, \alpha) \equiv (y - \alpha)^{\alpha - b} B_b^{(\alpha)}(y)$, (Bernoulli polynomial)

$$C(y, \alpha) = \frac{\alpha}{(y-\alpha)^2} .$$

(2) $g(y, \alpha) \equiv (1-y^2)^{\frac{\alpha+b+c+1}{2}} \left(\frac{1+y}{1-y}\right)^{\frac{\frac{1}{2}(c-b)(\alpha+b+c+1)}{2\alpha+b+c+2}} P_\alpha^{(b,c)}(y)$,

(Jacobi function)

$$C(y, \alpha) = \frac{-2(\alpha+1)(\alpha+b+c+1)}{(2\alpha+b+c+2)(1-y^2)^{3/2}} \left(\frac{1+y}{1-y}\right)^{\frac{\frac{1}{2}(c^2-b^2)}{(2\alpha+b+c+2)(2\alpha+b+c+4)}}$$

(3) $g(y, \alpha) \equiv (1-y^2)^{\frac{\alpha}{2}} \left(\frac{1+y}{1-y}\right)^{\frac{\frac{1}{2}(b-c)\alpha}{-2\alpha+b+c}} P_{-\alpha}^{(b,c)}(y)$,

$$C(y, \alpha) = \frac{2(\alpha-b)(\alpha-c)}{(2\alpha+b+c)(1-y^2)^{3/2}} \left(\frac{1+y}{1-y}\right)^{\frac{\frac{1}{2}(c^2-b^2)}{(-2\alpha+b+c-2)(-2\alpha+b+c)}}$$

$$(4)\qquad g(y,\alpha) = (y^2-1)^{\frac{\alpha+1}{2}} \left(\frac{1-y}{1+y}\right)^{\frac{b^2-c^2}{8(\alpha+1)}} ⓐ^{b,c}_{2\alpha}(y),$$

(Bateman's function)

$$C(y,\alpha) = \frac{(2\alpha-b-c+2)(2\alpha+b-c+2)}{4(\alpha+1)\ (y^2-1)^{3/2}} \left(\frac{1-y}{1+y}\right)^{\frac{b^2-c^2}{8(\alpha+1)(\alpha+2)}} .$$

$$(5)\qquad g(y,\alpha) = (y^2-1)^{\frac{\alpha}{2}} \left(\frac{1-y}{1+y}\right)^{\frac{b^2-c^2}{8\alpha}} ⓐ^{b,c}_{-2\alpha}(y),$$

$$C(y,\alpha) = \frac{(-2\alpha+b+c)(-2\alpha+b-c)}{4\alpha(y^2-1)^{3/2}} \left(\frac{1-y}{1+y}\right)^{\frac{b^2-c^2}{8\alpha(\alpha+1)}} .$$

These results were deduced from formulas (G.5), (D.3), (D.2), (D.7), (D.6), respectively.

In the special case when $b^2 = c^2$ the functions nos. 2, 3, 4, 5 above become reducible to solutions of the F-equation: v. solution no. 15, 17, 24, 25 in §4.

BIBLIOGRAPHY

APPELL, P.

1. Sur une classe de polynomes, Ann. École Norm., ser. 2, vol. 9 (1880), pp. 119-144.

BAILEY, W. N.

1. Generalized Hypergeometric Series, Cambridge Tracts in Mathematics and Mathematical Physics no. 32, Cambridge, 1935.

BARNES, E. W.

1. A new development of the theory of the hypergeometric functions, Proc. London Math. Soc., ser. 2, vol. 6 (1907), pp. 141-177.

BATEMAN, H.

1. A generalization of the Legendre polynomial, Proc. London Math. Soc., ser. 2, vol. 3 (1905), pp. 111-123.

2. Partial Differential Equations of Mathematical Physics, Cambridge, 1932.

3. Some simple differential difference equations and the related functions, Bull. Amer. Math. Soc., vol. 49 (1943), pp. 494-512.

BELL, E. T.

1. Exponential Polynomials, Annals of Math., vol. 35 (1934), pp. 258-277.

BELTRAMI, E.

1. Sulle funzione cilindriche, Atti Accad. Sci. Torino, vol. 16 (1880-1881), pp. 201-205.

2. Sulla teoria delle funzione potenziale simmetriche, Memorie dell' Accademia delle Scienze dell' Istituto di Bologna, ser. 4, vol. 2 (1880) pp. 461-505.

BRUWIER, L.

1. Sur une classe d'équations récurro-différentielles, Mémoires de la Societé Royale des Sciences de Liège, ser. 3, vol. 17 (1932), no. 22, pp. 1-48.

CAILLER, C.

1. Note sur une opération analytique, Mémoires de la Societé de Physique et d'Histoire Naturelle de Geneve, vol. 34 (1902-1905), pp. 296-368.

CALLANDREAU, M. O.

1. Sur le calcul des polynomes $X_n(\cos\theta)$ de Legendre pour les grandes valeurs de n, Bull. Sci. Math., vol. 15 (1891), pp. 121-124.

CHURCHILL, R. V.

1. Modern Operational Mathematics in Engineering, New York and London, 1944.

COPSON, E. T.

1. An Introduction to the Theory of Functions of a Complex Variable, Oxford, 1935.

CURZON, H. E. J.

1. On a connexion between the functions of Hermite and the functions of Legendre, Proc. London Math. Soc., ser. 2, vol. 12 (1912), pp. 236-259.

DERUYTS, J.

1. Sur une classe de polynomes analogues aux fonctions de Legendre, Mémoires de la Societé des Sciences de Liège, ser. 2, vol. 14 (1888), 2nd memoir.

DIDON, M.

1. Sur une équation aux dérivées partielles, Ann. École Norm., vol. 6 (1869), pp. 377-380.

DOETSCH, G.

1. Die in der Statistik seltener Ereignisse auftretenden Charlierschen Polynome und eine damit zusammenhangende Differential-differenzengleichung, Math. Ann., vol. 109 (1934), pp. 257-266.

2. Theorie und Anwendung der Laplace-Transformation. Berlin, 1937.

GEGENBAUER, L.

1. Über einige bestimmte Integrale, Akademie der Wissenschaften (Wien), Sitzungsberichte der mathematisch-naturwissenschaftlichen Classe, vol. 70, part 2 (1874), pp. 433-443.

2. Über eininge bestimmte Integrale, ibid., vol. 72, part 2 (1876), pp. 343-354.

GERONIMUS, J.

1. On the polynomials of Legendre and Hermite, Tohoku Math. J., vol. 34 (1931), pp. 295-296.

GLAISHER, J. W. L.

1. Notes on Laplace's coefficients, Proc. London Math. Soc., vol. 6 (1875), pp. 126-136.

HANKEL, H.

1. Die Euler'schen Integrale bei unbeschränkter Variabilität des Argumentes, Zeitschrift für Mathematik und Physik, vol. 9 (1864), pp. 1-21.

2. Die Cylinderfunctionen erster und zweiter Art, Math. Ann., vol. 1 (1869), pp. 467-501.

3. Bestimmte Integrale mit Cylinderfunctionen, Math. Ann., vol. 8 (1875), pp. 453-470.

HARDY, G. H.

1. Further researches in the theory of divergent series and integrals, Trans. Cambridge Philos. Soc., vol. 21 (1912), pp. 1-48.

2. Summation of a series of polynomials of Laguerre, J. London Math. Soc., vol. 7 (1932), pp. 138-139.

HEINE, E.

1. Handbuch der Kugelfunctionen, vol. 1, 2nd ed., Berlin, 1878.

HOBSON, E. W.

1. On Bessel's functions, and relations connecting them with hyperspherical and spherical harmonics, Proc. London Math. Soc., vol. 25 (1893), pp. 49-75.

2. The Theory of Spherical and Ellipsoidal Harmonics, Cambridge, 1931.

HUMBERT, P.

1. Sur les polynomes de Sonine à une et deux variables, J. École Polytech., ser. 2, vol. 24 (1924), pp. 59-75.

INCE, E. L.

1. Ordinary Differential Equations, London, 1927.

JACOBI, C. G. J.

1. Untersuchungen über die Differentialgleichung der hypergeometrischen Reihe, J. Reine Angew. Math., vol. 56 (1859), pp. 149-165; reprinted in Ges. Werke, vol. 6, pp. 184-201.

JONQUIERE, A.

1. Note sur la série $\sum_{n=1}^{\infty} x^n/n^s$, Bull. Soc. Math. France, vol. 17 (1889), pp. 142-152.

KNOPP, A.

1. Theory and Application of Infinite Series, London and Glasgow, 1928.

KOSHLIAKOV, N. S.

1. On Sonine's Polynomials, Mess. of Math., vol. 55 (1926), pp. 152-160.

KUMMER, E. E.

1. Über die hypergeometrische Reihe, J. Reine Angew. Math., vol. 15 (1836), pp. 39-83, 127-172.

de LAPLACE, P. S.

1. Oeuvres Complètes, vol. 5 (Mécanique Céleste, books 11-16), Paris, 1882.

LAURENT, M. H.

1. Mémoire sur les fonctions de Legendre, J. Math. Pures Appl., ser. 3, vol. 1 (1875), pp. 373-398.

LINDELOF, E.

1. Le Calcul des Résidus, Paris, 1905.

LIPSCHITZ, R.

1. Ueber ein Integral der Differentialgleichung $\partial^2 I/\partial x^2 + 1/x\, \partial I/\partial x + I = 0$, J. Reine Angew. Math., vol. 56 (1859), pp. 189-196.

LOMMEL, E.

1. Studien uber die Bessel'schen Functionen, Leipzig, 1868.

2. Zur Theorie der Bessel'schen Functionen, Math. Ann., vol. 3 (1871), pp. 475-487.

de LOUVILLE, J. E. d'A.

1. Éclaircissement sur une difficulté de statique proposée à l'academie, Histoire de l'Academie Royale des Sciences (année 1722), Paris, 1724, pp. 128-142.

MACDONALD, H. M.

1. Some applications of Fourier's theorem, Proc. London Math. Soc., vol. 35 (1902), pp. 428-433.

2. Note on the evaluation of certain integrals containing Bessel's functions, Proc. London Math. Soc., ser. 2, vol. 7 (1909), pp. 142-149.

MAGNUS, W. and OBERHETTINGER, F.

1. Formeln und Sätze für die speziellen Funktionen der mathematischen Physik, Berlin, 1943.

MILNE-THOMSON, L. M.

1. The Calculus of Finite Differences, London, 1933.

NÖRLUND, N. E.

1. Vorlesungen über Differenzenrechnung, Berlin, 1924.

2. Lecons sur les séries d'intérpolation, Paris, 1926.

PINCHERLE, S.

1. Della trasformazione di Laplace e di alcune sue applicazione, Memorie dell' Accademia delle Scienze dell' Istituto di Bologna, ser. 4, vol. 8 (1887), pp. 125-143.

RAINVILLE, E. D.

1. Notes on Legendre polynomials, Bull. Amer. Math. Soc., vol. 51 (1945), pp. 268-271.

SCHAFHEITLIN, P.

1. Über die Darstellung der hypergeometrischen Reihe durch ein bestimmtes Integrale, Math. Ann., vol. 30 (1887), pp. 157-158.

SCHLÄFLI, L.

1. Sopra un teorema di Jacobi recato a forma più generale ed applicato alla funzione cilindrica, Annali di Mat., ser. 2, vol. 5 (1873), pp. 199-205.

2. Über die zwei Heine'schen Kugelfunctionen mit beliebigem Parameter und ihre ausnahmlose Darstellung durch bestimmte Integrale, Bern, 1881.

SCHLÖMILCH, O.

1. Über die Bessel'sche Funktion, Zeitschrift für Mathematik und Physik, vol. 2 (1857), pp. 137-165.

SONINE, N.

1. Recherches sur les fonctions cylindriques et le développement des fonctions continues en séries, Math. Ann., vol. 16 (1880), pp. 1-80.

STRUVE, H.

1. Über den Einfluss der Diffraction an Fernröhren auf Lichtscheiben, Mémoires de l'Academie Impèriale des Sciences de St. Petersbourg, ser. 7, vol. 30, no. 8 (1882).

SZEGÖ, G.

1. Orthogonal Polynomials, New York, 1939.

TRUESDELL, C.

1. On a function which occurs in the theory of the structure of polymers, Ann. of Math., vol. 46 (1945), pp. 144-157.

2. On the functional equation $\partial/\partial z\ F(z, \alpha) = F(z, \alpha+1)$, Proc. Nat. Acad. Sci. U.S.A., vol. 33 (1947), pp. 82-93.

WATSON, G. N.

1. A Treatise on the Theory of Bessel Functions, 2nd ed., Cambridge, 1944.

2. Notes on generating functions of polynomials (1), J. London Math. Soc., vol. 7 (1932), pp. 138-139.

WEBER, H.

1. Über einige bestimmte Integrale, J. Reine Angew. Math., vol. 69 (1868), pp. 222-237.

2. Über die Bessel'schen Functionen und ihre Anwendung auf die Theorie der elektrischen Ströme, J. Reine Angew. Math., vol. 75 (1873), pp. 75-105.

WHITTAKER, E. T. and WATSON, G. N.

1. A Course of Modern Analysis, 4th ed., Cambridge, 1927.

WIRTINGER, W.

1. Über eine besondere Dirichletsche Reihe, J. Reine Angew. Math., vol. 129 (1905), pp. 214-219.

Milton Keynes UK
Ingram Content Group UK Ltd.
UKHW031527120324
439228UK00001B/46

9 780691 095776